Springer Series in

MATERIALS SCIENCE 108

Springer Series in
MATERIALS SCIENCE

Editors: R. Hull R. M. Osgood, Jr. J. Parisi H. Warlimont

The Springer Series in Materials Science covers the complete spectrum of materials physics, including fundamental principles, physical properties, materials theory and design. Recognizing the increasing importance of materials science in future device technologies, the book titles in this series reflect the state-of-the-art in understanding and controlling the structure and properties of all important classes of materials.

99 **Self-Organized Morphology in Nanostructured Materials**
Editors: K. Al-Shamery and J. Parisi

100 **Self Healing Materials**
An Alternative Approach to 20 Centuries of Materials Science
Editor: S. van der Zwaag

101 **New Organic Nanostructures for Next Generation Devices**
Editors: K. Al-Shamery, H.-G. Rubahn, and H. Sitter

102 **Photonic Crystal Fibers**
Properties and Applications
By F. Poli, A. Cucinotta, and S. Selleri

103 **Polarons in Advanced Materials**
Editor: A.S. Alexandrov

104 **Transparent Conductive Zinc Oxide**
Basics and Applications in Thin Film Solar Cells
Editors: K. Ellmer, A. Klein, and B. Rech

105 **Dilute III-V Nitride Semiconductors and Material Systems**
Physics and Technology
Editor: A. Erol

106 **Into The Nano Era**
Moore's Law Beyond Planar Silicon CMOS
Editor: H.R. Huff

107 **Organic Semiconductors in Sensor Applications**
Editors: D.A. Bernards, R.M. Ownes, and G.G. Malliaras

108 **Evolution of Thin Film Morphology**
Modeling and Simulations
By M. Pelliccione and T.-M. Lu

109 **Reactive Sputter Deposition**
Editors: D. Depla amd S. Mahieu

110 **The Physics of Organic Superconductors and Conductors**
Editor: A. Lebed

Volumes 50–98 are listed at the end of the book.

Matthew Pelliccione and Toh-Ming Lu

Evolution of Thin Film Morphology

Modeling and Simulations

Springer

Matthew Pelliccione
Department of Physics, Applied Physics and Astronomy, and Center for Integrated Electronics
Rensselaer Polytechnic Institute
Troy, NY 12180
USA

Toh-Ming Lu
Department of Physics, Applied Physics and Astronomy, and Center for Integrated Electronics
Rensselaer Polytechnic Institute
Troy, NY 12180
USA

Series Editors:

ISSN 0933-033X

ISBN: 978-0-387-75108-5 Springer Berlin Heidelberg New York
e-ISBN: 978-0-387-75109-2

Library of Congress Control Number: 2007940880

Springer is a part of Springer Science+Business Media.
springer.com

Typesetting: Data prepared by SPI Kolam using a Springer TEX macro package
Cover concept: eStudio Calamar Steinen
Cover production: WMX Design GmbH, Heidelberg

Printed on acid-free paper SPIN: 11559238 57/3180/SPI 5 4 3 2 1 0

Preface

Thin film deposition is the most ubiquitous and critical of the processes used to manufacture high-tech devices such as microprocessors, memories, solar cells, microelectromechanical systems (MEMS), lasers, solid-state lighting, and photovoltaics. The morphology and microstructure of thin films directly controls their optical, magnetic, and electrical properties, which are often significantly different from bulk material properties. Precise control of morphology and microstructure during thin film growth is paramount to producing the desired film quality for specific applications. To date, many thin film deposition techniques have been employed for manufacturing films, including thermal evaporation, sputter deposition, chemical vapor deposition, laser ablation, and electrochemical deposition.

The growth of films using these techniques often occurs under highly non-equilibrium conditions (sometimes referred to as far-from-equilibrium), which leads to a rough surface morphology and a complex temporal evolution. As atoms are deposited on a surface, atoms do not arrive at the surface at the same time uniformly across the surface. This random fluctuation, or noise, which is inherent to the deposition process, may create surface growth front roughness. The noise competes with surface smoothing processes, such as surface diffusion, to form a rough morphology if the experiment is performed at a sufficiently low temperature and / or at a high growth rate. In addition, growth front roughness can also be enhanced by growth processes such as geometrical shadowing. Due to the nature of the deposition process, atoms approaching the surface do not always approach in parallel; very often atoms arrive at the surface with an angular distribution. Therefore, some of the incident atoms will be captured at high points on a corrugated surface and may not reach the lower valleys of the surface, resulting in an enhancement of the growth front roughness. A conventional statistical mechanics treatment cannot be used to describe this complex growth phenomenon and as a result, the basic understanding of the dynamics of these systems relies very much on mathematical modeling and simulations.

The present monograph focuses on the modeling techniques used in research on morphology evolution during thin film growth. We emphasize the mathematical formulation of the problem in some detail both through numerical calculations based on Langevin continuum equations, and through Monte Carlo simulations based on discrete surface growth models when an analytical formulation is not convenient. In doing so, we follow the conceptual advancements made in understanding the morphological evolution of films during the last two and half decades. As such, we do not intend to include a comprehensive survey of the vast experimental works that have been reported in the literature.

An important milestone in the mathematical formulation used to describe the evolution of a growth front was presented more than two decades ago. This concept is based on a dynamic scaling hypothesis that utilizes an elegant model called self-affine scaling. Since then, numerous modeling, simulation, and experimental works have been reported based on dynamic scaling. Several books published recently have thoroughly discussed this subject, including *Fractal Concepts in Surface Growth* by A.-L. Barabási and H. E. Stanley (Cambridge University Press, 1995); and *Fractals, Scaling, and Growth Far from Equilibrium* by P. Meakin (Cambridge University Press, 1998). After the publication of these books, the field has grown considerably and the scope has broadened substantially. One of the salient developments is the recognition that films produced by common deposition techniques such as sputter deposition and chemical vapor deposition may not be self-affine, and have characteristics that have not been previously realized. Shadowing through a nonuniform flux distribution, for example, can profoundly affect the film morphology and lead to a breakdown of dynamic scaling. In addition to the common lateral correlation length scale, another length scale emerges called the wavelength that describes the distance between "mounds" that are formed under the shadowing effect. Also, the reemission effect, where incident atoms can "bounce around" before settling on the surface, can significantly change the surface morphology. Reemission is modeled with a sticking coefficient, which describes the probability that an atom "sticks" to the surface on impact. Depending on the value of the sticking coefficient, the morphology can change from a self-affine topology to a markedly different topology where the dynamic scaling hypothesis is no longer valid.

While following these conceptual developments on morphology evolution, the present monograph outlines the mathematical tools used to model these growth effects. The monograph is divided into three parts: Part I: Description of Thin Film Morphology, Part II: Continuum Surface Growth Models, and Part III: Discrete Surface Growth Models. In Part I, we introduce a set of useful statistics and correlation functions that have been utilized extensively in the literature to describe rough surfaces, including the root-mean-square roughness (interface width), lateral correlation length, autocorrelation function, height–height correlation function, and power spectral density function. Self-affine and non self-affine (mounded) surfaces are also introduced, as well as a discussion of the dynamic scaling hypothesis. In Part II, we outline how stochastic continuum equations are constructed to describe the evolution of growth front morphology, and explain the numerical methods that are used to solve these equations. We discuss both local models such as the random deposition model, Edwards–Wilkinson model, Mullins surface diffusion model, and the Kardar–Parisi–Zhang (KPZ) model, in addition to nonlocal models that include effects of shadowing and reemission. In particular, a connection between surface growth models with shadowing and reemission and a small world network model is discussed in detail. In Part III, discrete surface growth models based on Monte Carlo simulation techniques are introduced to describe the morphology evolution of thin films. Various aggregation strategies are described, including solid-on-solid techniques which are often used for relatively thin films, and ballistic aggregation techniques which are used to model thicker films. As an example, we use the results of these models, along with experimental results, to show the breakdown of dynamic scaling under common deposition conditions. Finally, the origin of a particular film impaction called "nodular defects" is discussed based on a ballistic aggregation model.

This monograph is useful for university researchers and industrial scientists working in the areas of semiconductor processing, optical coating, plasma etching, patterning, micromachining, polishing, tribology, and any discipline that requires an understanding of thin film growth processes. In particular, the reader is introduced to the mathematical tools that are available to describe such a complex problem, and lead to appreciate the utility of the various modeling methods through numerous example discussions. For beginners in the field, the text is written assuming a minimal background in mathematics and computer programming, which enables the readers to set up a computational program themselves to investigate specific topics of their interest in thin film deposition. Several of the simulations discussed in the text are implemented in the appendices to aid readers in creating their own growth models, and are also available on the Web at `http://www.stanford.edu/~pellim`.

MP was supported by the NSF IGERT program at Rensselaer. TML would like to thank Professor M. G. Lagally for his inspiration and encouragement over the years and long-time collaborator Professor G.-C. Wang for her tireless support. We thank our mentors and colleagues including Professors F. Family, J. G. Amar, R. van de Sanden, G. Palasantzas, J. D. Gunton, and G. Hong for invaluable discussions. Past collaborators including Dr. T. Karabacak, Dr. Y.-P. Zhao, Dr. J. T. Drotar, and Dr. H.-N. Yang have made many major contributions to the work discussed in this monograph.

Troy, NY
July, 2007

Matthew Pelliccione
Toh-Ming Lu

Contents

1

Introduction

The natural world is filled with rough surfaces. Roughness is, however, a relative term. One may describe a sheet of paper as being smooth to the touch, whereas on an atomic scale one would observe deep valleys and tall mountains in the landscape. Of particular scientific interest in the past few decades have been surfaces that exhibit this rough behavior on a nanometer scale, often referred to as thin film surfaces. Numerous studies have been carried out investigating processes to create thin films, characterize them, and test their physical properties [187]. The physics behind the growth and structure of these surfaces has been shown to be very interesting and challenging due to the complexities of the growth processes and surface structures [8, 40, 104, 112]. Specifically, surface and interface roughness controls many important physical and chemical properties of films. For example, the electrical conductivity of thin metal films depends very much on surface and interface roughness [135], and the reliability of a Si MOSFET (metal-oxide-semiconductor field-effect transistor) channel depends on the roughness of the gate oxide–silicon interface [82]. Also, interface roughness has a profound effect on the magnetic hysteresis of a magnetic film [115], and controls optical losses in optical waveguides [130]. Rough surfaces can increase the effective area for advanced charge storage devices [19], as well as promote capillary forces through wicking in modern heat pipe design [51]. These properties of thin films are exploited in a number of applications, including semiconductor devices [153], solar cells [127], and thin-film transistor (TFT) displays [73].

There are many different experimental methods for growing thin films in the lab, depending on the desired properties of the film. However, all methods accomplish the same general goal; to deposit matter on a substrate. Many deposition methods aim to deposit a specific type of material on a substrate, such as silicon, silicon dioxide, germanium, copper, or tantalum, but other compounds such as organic molecules can also be deposited. In order to create surfaces with nanometer scale roughness, the thickness of the deposited film is generally on the order of micrometers or nanometers, which means the surface must be grown layers of atoms at a time. To accomplish this, the material to

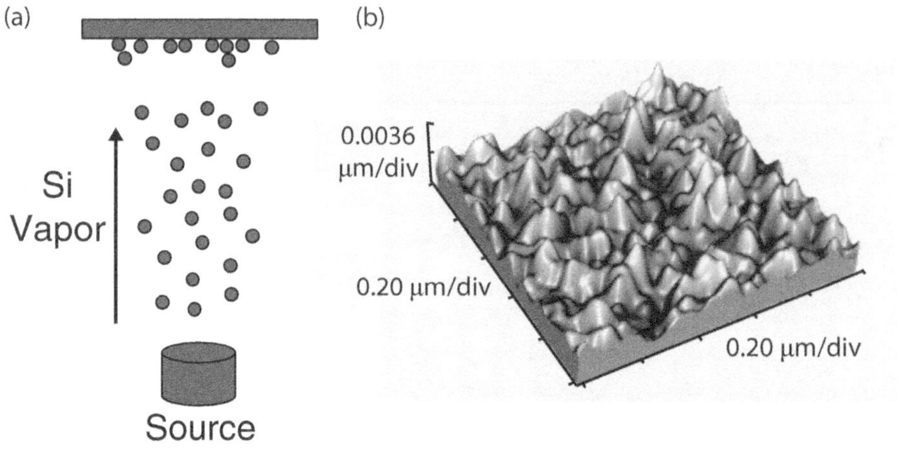

Fig. 1.1. (**a**) A schematic showing a thermal evaporation deposition experiment with a Si source. (**b**) An atomic force microscopy image of the surface morphology of a 2 μm thick amorphous Si film grown by thermal evaporation.

be deposited is often changed into a gaseous form in a vacuum to allow for atom-by-atom deposition on the surface.

The simplest deposition method is thermal evaporation [95], where the source material is placed in a crucible and then heated until it evaporates and condenses on a substrate located above the crucible. Figure 1.1a is a schematic drawing showing a thermal evaporation deposition experiment setup with a Si source. Figure 1.1b is an atomic force microscopy image of the surface morphology of a 2 μm thick amorphous Si film grown by the thermal evaporation technique at room temperature. As we can see from the image, the surface contains mountains and valleys over a certain length scale. The topology is obviously quite complex and it cannot be predicted deterministically. It belongs to a class of "complex phenomena" that has been pursued actively by scientists.

Once a thin film has been deposited, we need some way of quantitatively characterizing the surface. To this end, various mathematical tools have been developed that measure the most important properties of a surface, such as the mean height, roughness, and correlation length [187]. In addition, it has been found that many thin film surfaces obey certain common scaling properties that allow for a significant simplification of the description of the surface morphology. The most common such type of scaling is referred to as "self-affine" scaling, in which one can rescale the horizontal and vertical directions of the surface to obtain a new surface that is statistically identical to the original surface [100]. This definition of scaling is reminiscent of a fractal, and the mathematical concepts associated with fractals are used to describe self-affine surfaces. In particular, a self-affine surface is mainly characterized by a roughness exponent, which is related to the local roughness of the surface, but also the fractal dimension of the surface. A similar argument can be made about the scaling behavior of the surface profile in time, which is called "dynamic" scaling [8, 40, 41, 104]. Scaling arguments work quite well when the important growth effects in a deposition are "local", or only affect nearby surface heights, an example of which is surface diffusion, where atoms can diffuse to nearby locations depending on deposition conditions such as activation energy and temperature.

A problem arises when attempting to use self-affine scaling and dynamic scaling to describe thin film surfaces grown under the influence of nonlocal growth effects such as shadowing [123]. By definition, nonlocal growth effects are of much longer range than local effects, and as such are capable of defining a long-range length scale on the surface, often referred to as mound formation [122]. Mounds disrupt the self-affine behavior of the surface because they define a characteristic long-range length scale on the surface. When attempting to rescale the dimensions of the surface as in self-affine scaling, this characteristic length scale changes, and the rescaled surface is no longer statistically identical to the original surface. However, it has been shown that in growth processes that include only local growth effects mounded surfaces can be formed, as evidenced by surfaces created during molecular beam epitaxy [112, 166].

1.1 Growth Front Roughness

Many factors contribute to the formation of such a complex landscape on the surface of a film. First, there is always random noise that exists naturally during the deposition process because atoms do not arrive at the surface uniformly. These random fluctuations, which are inherent in the deposition process, can create growth front roughness. Noise competes with surface smoothing processes, such as surface diffusion, to form a rough morphology if the experiment is performed at either a sufficiently low temperature and / or at a high growth rate. In addition, growth front roughness can also be enhanced by growth processes such as geometrical shadowing. Shadowing is a result of deposition by a nonnormal incident flux [11, 62, 92, 106]. In many commonly employed deposition techniques such as sputter deposition [97, 144] and chemical vapor deposition [6, 31], atoms do not always approach the surface in parallel; very often they arrive at the surface with a distribution of trajectories. Figure 1.2 shows schematically the geometries of several commonly employed deposition techniques [92]. The angle θ is defined as the angle between the incident atomic flux and the surface normal. For conventional thermal evaporation or e-beam evaporation, if the substrate is sufficiently far away from the source and if the substrate dimensions are not too large, the flux arrives at the substrate with $\theta \approx 0°$, which is referred to as normal incidence. Oblique angle deposition can be achieved by tilting the substrate with respect to the particle flux in evaporation, and angles as large as $\theta \approx 85°$ are often used

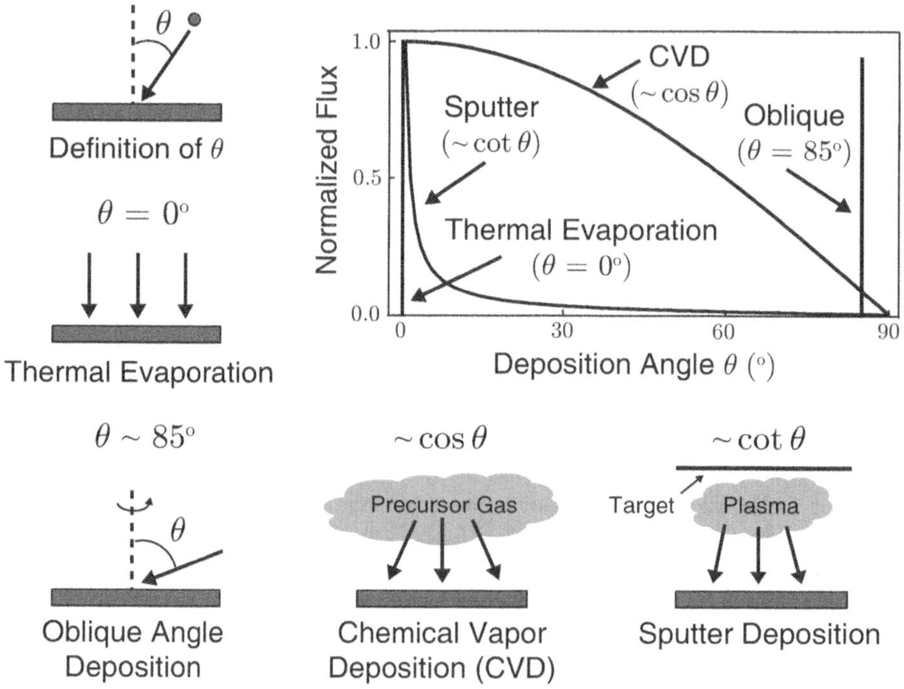

Fig. 1.2. Schematic diagrams showing the geometries of several commonly employed deposition techniques. The graph is a plot of the incident flux distribution of atoms arriving at the substrate for different deposition techniques. Depending on the geometry, sputter deposition can also be modeled with a cosine flux distribution [92].

experimentally [57, 76]. For chemical vapor deposition, precursor molecules may bounce around the deposition chamber numerous times before they undergo a reaction at the substrate. Therefore, the substrate experiences a molecular flux coming from a wide range of angles and can be represented by a cosine distribution. For sputter deposition, the distribution can be somewhat narrower (a ratio between cosine and sine functions) but, depending on the separation between the substrate and the source, can also be modeled by a cosine flux distribution. These nonnormal incident fluxes can lead to a shadowing effect during growth, as some of the incident atoms will be captured at high points on a corrugated surface at the expense of lower valleys on the surface, resulting in a dramatic enhancement of the growth front roughness.

Another important effect to consider is the value of the sticking coefficient [92, 184]. The sticking coefficient is defined as the probability that a particle will stick to the surface when it strikes. In both sputter deposition and chemical vapor deposition, the sticking coefficient may not be equal to unity. A nonunity sticking coefficient would allow the particle to be reemitted from

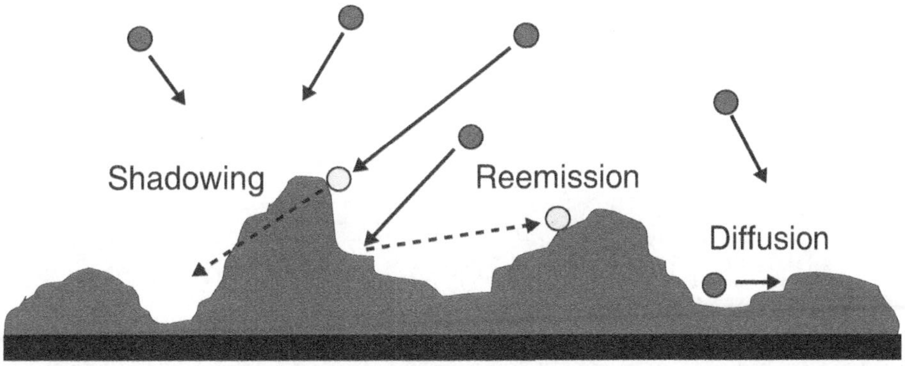

Fig. 1.3. Diagram of growth effects including diffusion, shadowing, and reemission that may affect surface morphology during thin film growth. The incident particle flux may arrive at the surface with a wide angular distribution depending on the deposition methods and parameters.

the surface upon impact. The particle may then deposit on the surface at a different location, or it may bounce around the surface more before it settles, which leads to a smoothing effect. Both shadowing and reemission effects are inherently nonlocal because an event that occurs at one place on the surface can affect the surface profile a far distance away. A summary of common growth effects is illustrated in Fig. 1.3.

1.2 Measurement Techniques

Before any analysis can be carried out regarding the roughness evolution of a surface, we must utilize measurement techniques that can reliably provide important information about a growth front. There are two classes of techniques that allow for a collection of quantitative information about the morphology of a growth front: real-space imaging techniques, and diffraction techniques [187]. Examples of real-space imaging techniques include atomic force microscopy (AFM), scanning tunneling microscopy (STM), scanning electron microscopy (SEM), and stylus profilometry. Real-space imaging techniques have the advantage of providing a direct visual interpretation of the surface morphology. From the surface profiles, one can extract all surface statistics relating to surface roughness. Examples of diffraction techniques include high-resolution low-energy electron diffraction (HRLEED), reflection high-energy electron diffraction (RHEED), atom diffraction, X-ray diffraction, and light scattering. For diffraction techniques, all surface roughness information can be extracted from the angular distribution of the diffracted radiation. Diffraction techniques have the advantages of providing noncontact measurements and the ability to obtain a statistical average of a large surface area in a short time. Also, some diffraction techniques are capable of, and many have the potential

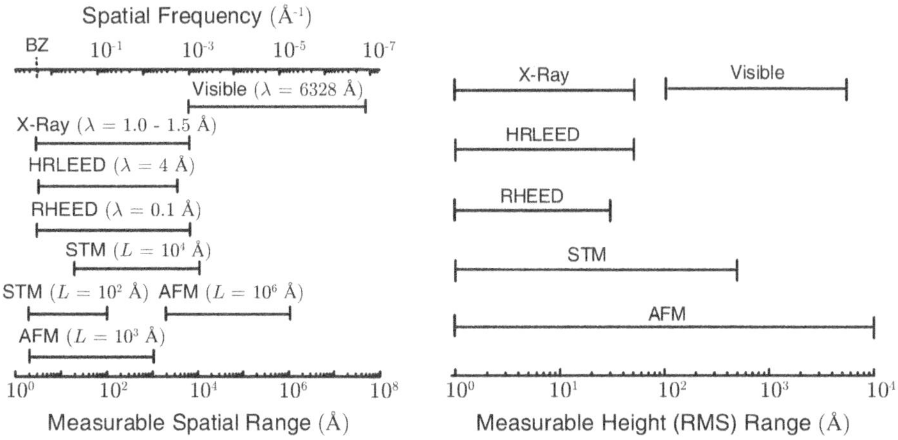

Fig. 1.4. Spatial length scale and frequency ranges for different imaging and diffraction techniques. For real-space imaging techniques, L represents the scan size, and for diffraction techniques, λ represents the wavelength of radiation used. The location of the Brillouin zone (BZ) is given for a lattice constant of approximately 2 Å, a length characteristic of experimental surfaces.

of, performing real-time measurements during growth or etching. Many of the real-space imaging and diffraction techniques are complementary to each other in the sense that they cover different length scales. Figure 1.4 shows a summary of the range of measurements each technique can cover in the lateral direction in terms of a spatial range, and the vertical direction in terms of the root-mean-square (RMS) roughness. More recently, it was shown that in situ spectroscopic ellipsometry can also provide useful information about the local surface roughness evolution [148]. Because the experimental characterization of growth front roughness is not the focus of this monograph, interested readers are referred to a recent book dedicated to this subject, *Characterization of Amorphous and Crystalline Rough Surface: Principles and Applications* by Y.-P. Zhao, G.-C. Wang, and T.-M. Lu (Academic Press, 2001).

1.3 Modeling

The main focus of this monograph is the modeling of thin film surface growth. Thin film growth models can be separated into two main categories: models that are based on continuum mathematics, and models that are based on discrete mathematics. In the past few decades, a number of models of both types have been proposed and have been shown to successfully predict properties of certain types of thin film growth, each with their own advantages and disadvantages. The ultimate utility of any of these theoretical models can be traced back to the core assumptions used to construct the model, which can be quite different for continuous and discrete models. Before discussing the specifics of any one particular model, it proves helpful to outline the basic assumptions of both types of models, which can then be used to judge the model validity when comparing to experimental results.

1.3.1 Continuum Models

Continuum models of thin film growth are often expressed as partial differential equations (PDEs) involving the surface height h at a position $\mathbf{x}$ on the substrate at time t. This PDE is often written as [8, 104]

$$\frac{\partial h(\mathbf{x},t)}{\partial t} = \Phi(h,\mathbf{x},t) + \eta(\mathbf{x},t), \tag{1.1}$$

where the term $\Phi(h,\mathbf{x},t)$ captures the growth effects to be modeled, and $\eta(\mathbf{x},t)$ represents the random noise inherent to the deposition. The noise is often chosen to be Gaussian noise because it is uncorrelated in space and time,

$$\langle \eta(\mathbf{x},t)\rangle = 0 \quad \text{and} \quad \langle \eta(\mathbf{x},t)\eta(\mathbf{x}',t')\rangle = 2D\delta^d(\mathbf{x}-\mathbf{x}')\delta(t-t'), \tag{1.2}$$

where d is the dimension of the vector $\mathbf{x}$, equal to the dimension of the substrate. Other types of correlated noise have been investigated as well, including power-law distributed noise [3, 77, 78, 93, 101, 109, 110, 125, 182, 183]. Without specifying the nature of the function Φ, we can make some statements about this type of model. First, due to the presence of noise, the model is not deterministic, and we should not be able to find an explicit solution $h(\mathbf{x},t)$. However, we may be able to predict average properties of $h(\mathbf{x},t)$, such as the mean surface height and surface roughness, without computing the analytical solution outright. As these statistics are averages over the entire domain of $\mathbf{x}$, the effects of noise will average out. As a result, when measuring the properties of an experimentally deposited thin film or the results of a model prediction, average statistics are all that is meaningful to compare because of random noise.

The primary advantage of a continuous growth model lies in the ability to choose the function Φ. Terms can be included in Φ depending on the type of growth effect one would like to model, which often come from considerations of how a certain growth effect would change the surface height profile $h(\mathbf{x},t)$. For example, it is shown in Chap. 5 that surface diffusion can be modeled by a term proportional to $-\nabla^4 h$ in Φ. The effect of surface diffusion can then be added or subtracted from a model easily by either including or excluding this term, leading to a concrete model where every term is included to model a specific growth effect. This then allows for a purely theoretical prediction of growth parameters used to characterize the surface.

Another advantage of continuum growth models lies in the concept of universality classes. These continuum models are able to predict values for growth parameters that can be experimentally measured, with specific theoretical predictions depending on the form of the function Φ. However, one finds that Φ often depends only on dominant growth effects in a deposition and not more specific conditions such as materials deposited, pressure, and temperature, which makes the prediction of the model rather general. As such, the specific theoretical predictions of growth parameters by a given form of Φ are said to form a universality class. For example, any deposition where surface diffusion is the only significant growth effect could be modeled by $\Phi \propto -\kappa\nabla^4 h$, with the predictions of the model valid for any deposition dominated by surface diffusion. In practice, often the concept of a universality class becomes less applicable because growth effects become much more complicated than can be modeled in this manner. It is common to measure growth parameters in a somewhat continuous fashion as opposed to only observing values for growth parameters in discrete sets as a universality class would suggest. Even so, for relatively simple growth effects, this concept is useful to deduce dominant growth effects from continuous modeling by comparing predictions of different forms of Φ with experimental results.

Although these continuous models allow considerable freedom in choosing the dominant growth effects to model, along with requiring a certain amount of creativity to derive the form of terms in Φ, it comes at the cost of utility. For realistic surfaces, these continuum PDEs can become quite complex, usually involving nonlinear terms, which poses a problem when one attempts to solve these models numerically from both an accuracy and efficiency standpoint. (In practice, continuum models must be numerically integrated under specific boundary conditions to give testable results. Although numerical integration itself requires a discretization of the continuous problem, these are not considered "discrete" models, which are discussed in the next section, as they are based on continuum mathematics.) In addition, from the assumption that the surface can be modeled by a height function $h(\mathbf{x}, t)$, any model prediction of this type will necessarily yield a surface where each position $\mathbf{x}$ corresponds to one surface height because $h(\mathbf{x}, t)$ is a function. It is also possible to have surfaces with overhangs, which would require the use of a multi-valued function to describe the surface. Care must be taken in applying these continuum models to depositions that may yield surfaces with overhangs.

1.3.2 Discrete Models

Discrete growth models offer an alternative method with which to model thin film growth that alleviates some of the problems encountered with continuous growth models. Discrete models are attractive from a modeling perspective because they are relatively simple to create and can quickly yield tangible predictions. In particular, these models evolve the system under a set of simple rules that can lead to complex behavior. These rules are stochastic in nature, leading to the requirement that averages must be taken to obtain results comparable to experiments. However, because of their relative simplicity, it is often possible to perform numerous runs of a discrete model to take such an average.

The most common type of discrete simulation used in thin film growth modeling is called Monte Carlo (MC), so named because of the randomness inherent to the algorithms. The term "Monte Carlo" is widely used in physical modeling [42], with applications in finance, chemistry, and particle physics to name a few, and in each discipline may be implemented somewhat differently. In the context of thin film growth, MC models tend to be models that examine general morphological behavior, often ignoring details such as the specific types of atoms being deposited, or the specific nature of interatomic forces. Models that include this level of detail are often referred to as molecular dynamics (MD) simulations. We will not discuss MD methods in this monograph, the interested reader is referred to the available literature on the subject [5, 133]. Even though MC simulations ignore specific details of a deposition, MC methods are able to provide a significant amount of information regarding the evolution of a growth front, and are a focus of this monograph.

In a sense, these discrete simulations are a combination of theoretical and experimental techniques. Often, there is no complete analytical theory upon which these models reside because of the complexity that would be required in such a theory. Predictions made by these models are based on data analysis and observation, as would be the case in an experiment, the only difference being the arena in which the measurements are taken. The theoretical aspect of the models lies in choosing the effects to include in the simulation, and determining how those effects would manifest themselves in the system to be modeled. Herein lies a tremendous advantage of these discrete models, the ability to pick and choose what growth effects to model and the ability to observe the effects of such a choice with relative ease. For example, if one wanted to investigate the behavior of a growth front when diffusion is negligible, experimentally one would have to ensure that the temperature during a deposition is low enough, or that non-diffusive materials are used in an experiment. In these models, one would simply turn off the diffusion effects in the simulation code to observe the effects of low diffusion, which can save time and effort as compared to a purely experimental investigation.

However, this freedom in choosing growth effects can also be a disadvantage because results of such a model can be somewhat "artificial" if the model assumptions do not closely mimic experimental conditions. Especially when creating and testing a model from scratch, one often has an idea of what a model should reasonably give as a result, but one must ensure that any new observations given by the model are truly due to the physics of the problem, and not an artifact of poor model assumptions or simply a bug in the simulation code. As such, there is a danger in constructing a model only after all experimental data have been taken. In constructing the model, one knows what the outcome "should" be, and it is tempting to create a model that agrees with experimental data and claim that the model is correct. It is possible, however, that different models would also be consistent with current data, and claiming that one particular model is superior would be up for debate. This problem is remedied by using these models to predict the results of a new experiment, one whose outcome was unknown when the models were formulated. Unfortunately, it may not always be feasible to conduct new experiments that test differences in models. Even so, new physics is often first observed experimentally and then incorporated into theoretical models, and it is up to the individual to decide, based on the model assumptions, the validity of the model.

Part I

Description of Thin Film Morphology

2

Surface Statistics

Mathematically, rough surfaces can be described by a surface height profile $h(\mathbf{x}, t)$, where h denotes the surface height with respect to the substrate at a position $\mathbf{x}$ on the surface at time t. The functional form of $h(\mathbf{x}, t)$ implies that there is only one surface height at position $\mathbf{x}$, which may not hold for surfaces with overhangs. In the discussion that follows, the surface height profile is assumed to be a single-valued function.

To define the various statistics used to characterize rough surfaces, it is convenient to define the concept of an average in this context. The average of a function $f(\mathbf{x}, t)$, denoted as $\langle f(\mathbf{x}, t)\rangle$, is defined as

$$\langle f(\mathbf{x}, t)\rangle \equiv \frac{\int f(\mathbf{x}, t)\mathrm{d}\mathbf{x}}{\int \mathrm{d}\mathbf{x}}, \tag{2.1}$$

where the domain of integration is the domain of the d-dimensional substrate, and the vector $\mathbf{x}$ is d-dimensional. Surface growth is commonly referred to as taking place in "d+1" dimensions, which means that the substrate is d-dimensional, and the growth takes place in one extra dimension. For example, growth on a two-dimensional substrate occurs in three dimensions because the vertical growth of the surface occurs normal to the substrate. Growth in 2+1 dimensions is the most common experimentally as depositions usually occur on a planar substrate. As such, we concentrate our discussion primarily on 2+1 dimensions, although theoretical results are given for a general dimension d. It is noted that the general mathematical definition of average includes a probability density function $P(\mathbf{x}, t)$ in the integrand. However, because all surface heights are to be weighted equally, $P(\mathbf{x}, t)$ is constant in the domain of integration and zero outside the domain, which is consistent with (2.1). Also, if the domain is discrete rather than continuous, the integral can be replaced by a discrete summation of all surface points. Measuring statistics from a discrete surface is discussed in Sect. 2.9.

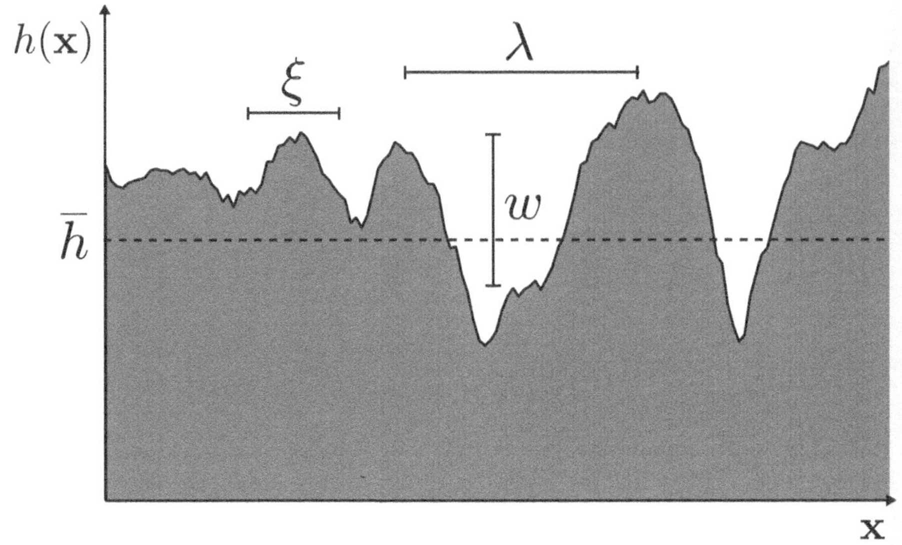

Fig. 2.1. Illustration of statistics used to describe rough surfaces. The definitions of the mean height $\overline{h}$, interface width w, lateral correlation length ξ, and wavelength λ are given in this chapter.

2.1 Mean Height

The mean height $\overline{h}$ of a surface profile is defined as

$$\overline{h}(t) \equiv \langle h(\mathbf{x}, t) \rangle. \tag{2.2}$$

It is very common to redefine the surface height profile such that $\overline{h} = 0$ by choosing a suitable reference height. This is helpful when concentrating on surface height fluctuations because any artificial effect introduced by the mean height is removed. In the definitions that follow, the mean height is taken to be equal to zero at all times. To obtain the definition for a surface with a nonzero mean height, simply replace $h(\mathbf{x}, t)$ with $(h(\mathbf{x}, t) - \overline{h})$. If the reference height remains constant in time with respect to the substrate, and if the flux of particles deposited on the surface is uniform in time, the mean height will be linear in time, $\overline{h} \sim t$, because the mean height is proportional to the total number of particles deposited on the surface.

2.2 Interface Width

The most common statistic used to describe the roughness of a surface is the standard deviation w of the surface heights, also called the interface width or root-mean-square (RMS) roughness. The interface width is defined as

$$w(t) \equiv \sqrt{\left\langle [h(\mathbf{x},t)]^2 \right\rangle}. \tag{2.3}$$

Larger values of the interface width indicate a rougher surface. It is common to observe a power-law behavior for the interface width in deposition time,

$$w(t) \sim t^{\beta}, \tag{2.4}$$

where β is referred to as the growth exponent. This characteristic behavior of the interface width is the basis for dynamic scaling theory, which has been widely used to describe the dynamic properties of thin films.

2.3 Autocorrelation Function

Statistics such as the mean height and interface width measure the vertical properties of a surface and do not reflect correlations between different lateral positions on the surface. To accomplish this, the autocorrelation function $R(\mathbf{r}, t)$ is introduced, which measures the correlation of surface heights separated laterally by the vector $\mathbf{r}$. The autocorrelation function is defined as

$$R(\mathbf{r}, t) \equiv w^{-2} \langle h(\mathbf{x}, t) h(\mathbf{x} + \mathbf{r}, t) \rangle. \tag{2.5}$$

If the statistical behavior of a surface does not depend on the specific orientation of the surface, the surface is said to be isotropic, and the autocorrelation function depends only on $|\mathbf{r}|$. Thus, a new variable $r = |\mathbf{r}|$ can be introduced to express the autocorrelation function as $R(r, t)$. Surfaces that do not possess this symmetry are called anistropic surfaces, whose treatment is not discussed here. The interested reader may find more information about anistropic surfaces in [187].

General properties of the autocorrelation function can be deduced from its definition. When $r = 0$, $R(0, t) = 1$, using the definition of interface width to evaluate the average. In addition, when r is large, surface heights become uncorrelated. Because $\langle xy \rangle = \langle x \rangle \langle y \rangle$ if x and y are uncorrelated variables, for large r,

$$R(r, t) \rightarrow w^{-2} \langle h(\mathbf{x}, t) \rangle \langle h(\mathbf{x} + \mathbf{r}, t) \rangle \sim w^{-2} \left(\overline{h}\right)^2 \sim 0, \tag{2.6}$$

as the mean height $\overline{h}$ is taken to be zero at all times by the choice of reference height. It follows that $R(r, t)$ is a decreasing function of r, and how fast $R(r, t)$ decreases is a measure of the lateral correlation of surface heights. For self-affine thin film surfaces, the autocorrelation function is often found to have an exponentially decreasing behavior, which naturally satisfies the above properties. Mounded thin film surfaces also exhibit a decreasing autocorrelation function in general, but the autocorrelation function may also exhibit oscillations as a result of the presence of mounds. Figure 2.2a shows the characteristic behavior of the autocorrelation function of a self-affine rough surface.

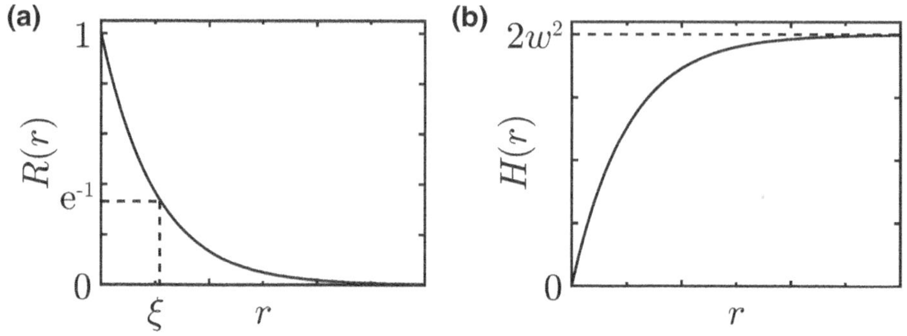

Fig. 2.2. Plot of the general behavior of the (**a**) autocorrelation function and (**b**) height–height correlation function for a self-affine rough surface. From (2.10), the height–height correlation function is simply an inversion and vertical translation of the autocorrelation function.

2.4 Lateral Correlation Length

Motivated by the properties of the autocorrelation function, the lateral correlation length ξ is defined as the value of r at which $R(r,t)$ decreases to $1/\mathrm{e}$ of its original value,

$$R(\xi, t) \equiv \mathrm{e}^{-1}. \tag{2.7}$$

It follows that two surface heights are significantly correlated on average if their lateral separation is less than the lateral correlation length ξ. In some contexts, continuum models for the autocorrelation function are used to define the correlation length that may differ from this definition. Regardless of the specific definition, the correlation length must be measured in a consistent manner to be meaningful. The time-dependent behavior of the lateral correlation length is often found to be a power law,

$$\xi(t) \sim t^{1/z}, \tag{2.8}$$

where z is referred to as the dynamic exponent.

2.5 Height–Height Correlation Function

A similar correlation function commonly used in scaling arguments is the height–height correlation function $H(\mathbf{r}, t)$ defined as

$$H(\mathbf{r}, t) \equiv \left\langle (h(\mathbf{x}+\mathbf{r}, t) - h(\mathbf{x}, t))^2 \right\rangle . \tag{2.9}$$

The properties of $H(\mathbf{r}, t)$ can be inferred from the properties of the autocorrelation function from the relation

$$
\begin{aligned}
H(\mathbf{r},t) &= \left\langle (h(\mathbf{x}+\mathbf{r},t) - h(\mathbf{x},t))^2 \right\rangle \\
&= \left\langle [h(\mathbf{x}+\mathbf{r},t)]^2 \right\rangle + \left\langle [h(\mathbf{x},t)]^2 \right\rangle - 2\left\langle h(\mathbf{x}+\mathbf{r},t)h(\mathbf{x},t)\right\rangle \\
&= 2w^2 - 2\left(w^2 R(\mathbf{r},t)\right) \\
&= 2w^2\left[1 - R(\mathbf{r},t)\right]. \qquad (2.10)
\end{aligned}
$$

From the properties of the autocorrelation function $R(\mathbf{r},t)$, it follows that $H(\mathbf{0},t) = 0$ and $H(r,t) \sim 2w^2$ for $r \gg \xi$. This behavior is seen in Fig. 2.2b. As with the autocorrelation function, the height–height correlation function is a function of $r = |\mathbf{r}|$ only for isotropic surfaces, which allows the height–height correlation function to be expressed as $H(r,t)$. The usefulness of this new correlation function is in the behavior of the function for small r. For most thin film surfaces, the height–height correlation function behaves as a power law for small r, which obeys certain scaling properties as discussed in Sect. 2.8.1.

2.6 Root-Mean-Square (RMS) Surface Slope

The root-mean-square surface slope ϱ is defined as

$$\varrho \equiv \sqrt{\langle |\nabla h(\mathbf{x},t)|^2 \rangle}. \qquad (2.11)$$

Using integration by parts, (2.11) can be represented as

$$
\begin{aligned}
\varrho^2 &= \int \nabla h(\mathbf{x}) \cdot \nabla h(\mathbf{x}) \mathrm{d}\mathbf{x} \\
&= \int h(\mathbf{x}) \frac{\partial h(\mathbf{x})}{\partial n} \mathrm{d}S - \int h(\mathbf{x}) \nabla^2 h(\mathbf{x}) \mathrm{d}\mathbf{x}.
\end{aligned}
$$

If the mean is taken to be zero at all times and the surface is isotropic, the surface integral will average out to zero over a sufficiently large domain. If the surface integral can be neglected, we find

$$
\begin{aligned}
\varrho^2 &= -\int h(\mathbf{x}) \nabla^2 h(\mathbf{x}) \mathrm{d}\mathbf{x} \\
&= -\int h(\mathbf{x}) \left[\nabla^2_{\mathbf{r}} h(\mathbf{x}+\mathbf{r})\right]_{\mathbf{r}=\mathbf{0}} \mathrm{d}\mathbf{x} \\
&= -\nabla^2_{\mathbf{r}} \left[\int h(\mathbf{x}) h(\mathbf{x}+\mathbf{r}) \mathrm{d}\mathbf{x}\right]_{\mathbf{r}=\mathbf{0}} \\
&= -w^2 \nabla^2_{\mathbf{r}} R(\mathbf{r}=\mathbf{0}). \qquad (2.12)
\end{aligned}
$$

In 1+1 dimensions, if the autocorrelation function is a function of $r = |\mathbf{r}|$ only, this relation becomes

$$\varrho^2 = -w^2 \left[\frac{\mathrm{d}^2 R(r)}{\mathrm{d}r^2} \right]_{r=0} . \qquad (2.13)$$

In 2+1 dimensions, if the autocorrelation function is a function of $r = |\mathbf{r}|$ only, this relation becomes

$$\varrho^2 = -w^2 \left[\left(\frac{\mathrm{d}^2}{\mathrm{d}r^2} + \frac{1}{r}\frac{\mathrm{d}}{\mathrm{d}r} \right) R(r) \right]_{r=0} . \qquad (2.14)$$

Applying this formula to a self-affine surface may lead to a divergence. The local slope m can be introduced to remedy this problem; further discussion is given in Sect. 3.3.

2.7 Power Spectral Density Function

The lateral correlation length represents the short-range lateral behavior of a surface, but beyond the lateral correlation length, even though surface heights are not significantly correlated, they may exhibit a periodic behavior on a length scale larger than the lateral correlation length. In order to determine this long-range behavior, the power spectral density function (PSD), also known as the structure function, is used. The PSD is related to a d-dimensional Fourier transform of the surface heights, defined in reciprocal space as

$$P(\mathbf{k}, t) \equiv \frac{1}{(2\pi)^d} \left| \langle h(\mathbf{x}, t) \mathrm{e}^{-\mathrm{i}\mathbf{k}\cdot\mathbf{x}} \rangle \right|^2 . \qquad (2.15)$$

To avoid confusion, we note that some authors use the variable $\mathbf{q}$ instead of $\mathbf{k}$ in the definition of the PSD. To obtain an alternate representation of $P(\mathbf{k}, t)$, expand (2.15) to give

$$\begin{aligned} P(\mathbf{k}, t) &= \frac{1}{(2\pi)^d} \left[\int h(\mathbf{x}, t) \mathrm{e}^{-\mathrm{i}\mathbf{k}\cdot\mathbf{x}} \mathrm{d}\mathbf{x} \right] \left[\int h(\mathbf{x}', t) \mathrm{e}^{\mathrm{i}\mathbf{k}\cdot\mathbf{x}'} \mathrm{d}\mathbf{x}' \right] \\ &= \frac{1}{(2\pi)^d} \int\int h(\mathbf{x}, t) h(\mathbf{x}', t) \mathrm{e}^{-\mathrm{i}\mathbf{k}\cdot(\mathbf{x}-\mathbf{x}')} \mathrm{d}\mathbf{x}\mathrm{d}\mathbf{x}' \\ &= \frac{1}{(2\pi)^d} \int\int h(\mathbf{r}, t) h(\mathbf{r}+\mathbf{r}', t) \mathrm{e}^{\mathrm{i}\mathbf{k}\cdot\mathbf{r}'} \mathrm{d}\mathbf{r}\mathrm{d}\mathbf{r}' \\ &= \frac{1}{(2\pi)^d} \int \left[\int h(\mathbf{r}, t) h(\mathbf{r}+\mathbf{r}', t) \mathrm{d}\mathbf{r} \right] \mathrm{e}^{\mathrm{i}\mathbf{k}\cdot\mathbf{r}'} \mathrm{d}\mathbf{r}', \end{aligned}$$

where the change of variables $\mathbf{x} = \mathbf{r}$ and $\mathbf{x}' = \mathbf{r} + \mathbf{r}'$ was used and the integration is over the entire domain of $\mathbf{r}$ and $\mathbf{r}'$. Taking advantage of the definition of the autocorrelation function from (2.5),

$$P(\mathbf{k}, t) = \frac{w^2}{(2\pi)^d} \int R(\mathbf{r}, t) \mathrm{e}^{\mathrm{i}\mathbf{k}\cdot\mathbf{r}} \mathrm{d}\mathbf{r}. \qquad (2.16)$$

The power spectral density function is a Fourier transform of the autocorrelation function. Using this definition, the total "area" in **k**-space enclosed by the PSD is equal to w^2,

$$\begin{aligned}\int P(\mathbf{k},t)\mathrm{d}\mathbf{k} &= \frac{w^2}{(2\pi)^d}\int R(\mathbf{r},t)\mathrm{d}\mathbf{r}\int \mathrm{e}^{\mathrm{i}\mathbf{k}\cdot\mathbf{r}}\mathrm{d}\mathbf{k}\\ &= \frac{w^2}{(2\pi)^d}\int R(\mathbf{r},t)\mathrm{d}\mathbf{r}\left[(2\pi)^d\delta^d(\mathbf{r})\right]\\ &= w^2R(\mathbf{0},t) = w^2,\end{aligned} \tag{2.17}$$

because $R(\mathbf{0},t) = 1$ by definition. The PSD can also be expressed in terms of the height–height correlation function as

$$P(\mathbf{k},t) = \frac{1}{2(2\pi)^d}\int \left[2w^2 - H(\mathbf{r},t)\right]\mathrm{e}^{\mathrm{i}\mathbf{k}\cdot\mathbf{r}}\mathrm{d}\mathbf{r}. \tag{2.18}$$

To find the PSD of a 1+1 dimensional surface, take $d = 1$ and express the PSD as

$$P(k,t) = \frac{w^2}{\pi}\int_0^\infty R(r,t)\cos(kr)\mathrm{d}r, \tag{2.19}$$

which follows because $R(r)$ is even. To find the PSD of an isotropic 2+1 dimensional surface, take $d = 2$ and $\mathbf{k}\cdot\mathbf{r} = kr\cos\theta$ in polar coordinates. For isotropic surfaces, $P(\mathbf{k},t)$ depends only on $k = |\mathbf{k}|$, so the PSD can be expressed as $P(k,t)$. The PSD can then be written as

$$P(k,t) = \frac{w^2}{(2\pi)^2}\int_0^{2\pi}\int_0^\infty R(r,t)\mathrm{e}^{ikr\cos\theta}r\mathrm{d}r\mathrm{d}\theta.$$

The angular integral can be evaluated to give

$$\int_0^{2\pi}\mathrm{e}^{ikr\cos\theta}\mathrm{d}\theta = 2\pi J_0(kr),$$

where $J_0(x)$ is the zeroth-order Bessel function as discussed in Sect. A.1.1. It follows that

$$P(k,t) = \frac{w^2}{2\pi}\int_0^\infty R(r,t)rJ_0(kr)\mathrm{d}r, \tag{2.20}$$

for an isotropic 2+1 dimensional surface. This form can be simplified by the definition of a Hankel transform $\mathcal{H}$,

$$\mathcal{H}\left\{R(r,t)\right\} \equiv \int_0^\infty R(r,t)rJ_0(kr)\mathrm{d}r, \tag{2.21}$$

which is discussed in Sect. A.3. The PSD then becomes

$$P(k,t) = \frac{w^2}{2\pi}\mathcal{H}\left\{R(r,t)\right\}. \tag{2.22}$$

If the PSD spectrum exhibits a characteristic peak at a wavenumber k_m, the surface possesses a long-range periodic behavior and is said to exhibit wavelength selection at a wavelength

$$\lambda \equiv 2\pi k_m^{-1}. \tag{2.23}$$

Surfaces that exhibit wavelength selection are said to be mounded. If the PSD exhibits a peak, the peak position k_m generally has a power-law behavior in time,

$$k_m(t) \sim t^{-p}, \tag{2.24}$$

where p is referred to as the wavelength exponent. Representative plots of the PSD for different types of surfaces can be found in Fig. 3.4 on p. 40 and Fig. 4.5 on p. 54. In Fig. 3.4, the surface does not exhibit wavelength selection because the PSD has no characteristic peak, whereas in Fig. 4.5 a peak is clearly seen, which indicates that the surface is mounded.

2.8 Scaling

The concept of scaling is a powerful tool that allows for a considerable simplification of the description of thin film rough surfaces. Scaling is often described in terms of a scaling function that describes certain aspects of a rough surface. There are two main types of scaling functions in this context: functions that are invariant under scale transformations and functions that do not change their characteristic behavior in time. These scaling concepts are very similar, and are often both referred to simply as scaling. However, there are some key differences in the behavior of these types of scaling and the physical results they imply.

2.8.1 Self-Affine Scaling

Let a function $f(x_1, x_2, \ldots, x_n)$ be a function of n variables x_i, for example, the surface height profile of a rough surface at some time t_0. The function f is said to exhibit self-affine scaling [100] if, for some function $g(\varepsilon_1, \varepsilon_2, \ldots, \varepsilon_n)$,

$$f(\varepsilon_1 x_1, \varepsilon_2 x_2, \ldots, \varepsilon_n x_n) = g(\varepsilon_1, \varepsilon_2, \ldots, \varepsilon_n) f(x_1, x_2, \ldots, x_n). \tag{2.25}$$

This definition implies that if the variable x_i has been rescaled by a factor ε_i, the resulting function is a constant factor multiplied by the original function. Note that this notion of scaling is a property of the function f only, and is in no means related to the behavior of f at other times t. From (2.25), for a single-variable function $f(x)$,

$$\begin{aligned} f(abx) &= g(ab)f(x); \\ f(abx) &= g(a)f(bx) \\ &= g(a)g(b)f(x). \end{aligned}$$

Therefore,

$$g(ab) = g(a)g(b). \tag{2.26}$$

To determine a functional form for g, assume it possesses a continuous first derivative and differentiate (2.26) with respect to a to obtain

$$b\frac{\mathrm{d}g(ab)}{\mathrm{d}(ab)} = \frac{\mathrm{d}g(a)}{\mathrm{d}a}g(b).$$

Because this holds for all a and b, evaluate the previous expression at $a = 1$. Rearranging terms gives, with $[\mathrm{d}g(a)/\mathrm{d}a]_{a=1} = g'(1)$,

$$\frac{\mathrm{d}g(b)}{g(b)} = g'(1)\frac{\mathrm{d}b}{b}.$$

Integrating yields

$$g(b) = b^k, \tag{2.27}$$

where $k = g'(1)$ and $g(1) = 1$ from the definition of g. Thus, if a function f exhibits self-affine scaling, the function values scale as a power law. Using this property of self-affine scaling functions, self-affine rough surfaces will be defined, which form the basis for dynamic scaling theory and the general description of thin film rough surfaces.

Note that in the above derivation, the function f was not assumed to be a smooth function of its arguments. If we further assume that f has a continuous first derivative, we can obtain a closed-form expression for f. For simplicity, assume f is a function of one variable x. Then

$$f(\varepsilon x) = g(\varepsilon)f(x). \tag{2.28}$$

However, since $f(\varepsilon x) = f(x\varepsilon)$, it follows that

$$g(\varepsilon)f(x) = g(x)f(\varepsilon),$$

or, rearranging terms,

$$f(x) = g(x)\left(\frac{f(\varepsilon)}{g(\varepsilon)}\right),$$

for $g(\varepsilon) \neq 0$. However, from (2.28) with $x = 1$,

$$f(\varepsilon) = g(\varepsilon)f(1).$$

Substituting, this yields

$$f(x) = g(x)\left(\frac{g(\varepsilon)f(1)}{g(\varepsilon)}\right) = g(x)f(1).$$

If f is assumed to have a continuous first derivative, then, because $g(x) = x^k$ for smooth g from (2.27), f must take the form

$$f(x) = cx^k. \tag{2.29}$$

This result states that any smooth function f that exhibits self-affine scaling is a power law. Experimentally, surface height profiles need not be smooth functions of position because the surface must be discretized to measure the profile. One such example is given in Fig. 2.1, which is a surface profile obtained experimentally by atomic force microscopy. The derivative is not continuous through the kinks in this profile. In addition, real surfaces that exhibit self-affine behavior cannot do so to arbitrarily small length scales, as the surface profile is not well defined at length scales smaller than the size of an atom. Thus, surface height profiles that exhibit self-affine scaling need not be a power law as in (2.29). However, lateral correlation functions are smooth, and a power-law behavior for a lateral correlation function may imply a self-affine scaling behavior for the surface.

A function f is said to exhibit self-similar scaling if $g(\varepsilon) = \varepsilon$ in (2.28). Conceptually, this means that rescaling the arguments of f and the value of f by the same factor yields the original function f. In this respect, self-similar functions are a special case of self-affine functions. Figure 2.3a is an example of a self-similar function. From the previous discussion, the only class of smooth self-similar functions is $f(x) = cx$. Figure 2.3b is an example of a self-affine function, $f(x) = \sqrt{x}$, where $g(\varepsilon) = \sqrt{\varepsilon}$. Note that the scale factors in the vertical and horizontal directions are equal for a self-similar function, but not in general for a self-affine function.

2.8.2 Time-Dependent Scaling

Consider a function $F(x,t)$ that explicitly includes time as an independent variable. This function is said to exhibit time-dependent scaling if there exist two functions $s_1(t)$ and $s_2(t)$ such that

$$\frac{\partial}{\partial t}\left[s_1(t)F(s_2(t)x,t)\right] = 0, \tag{2.30}$$

or, equivalently,

$$s_1(t)F(s_2(t)x,t) = G(x), \tag{2.31}$$

where $G(x)$ is independent of time. This definition implies that if $F(x,t)$ is graphed versus x separately at times $t_1, t_2, t_3, \ldots$, and the axes of each graph are rescaled by the appropriate factors given by $s_{1,2}(t)$, the same curve will be obtained in each scaled graph. Note that this notion of scaling is different than the self-affine scaling described in Sect. 2.8.1. In self-affine scaling, scale factors are used to relate the function back to itself. Time-dependent scaling uses scale factors to eliminate one of the independent variables of the function.

As an example, let us consider the function

$$H(r,t) = 2t^{2\beta}\left[1 - \exp\left(-\frac{r^{2\alpha}}{t^{2\beta}}\right)\right]. \tag{2.32}$$

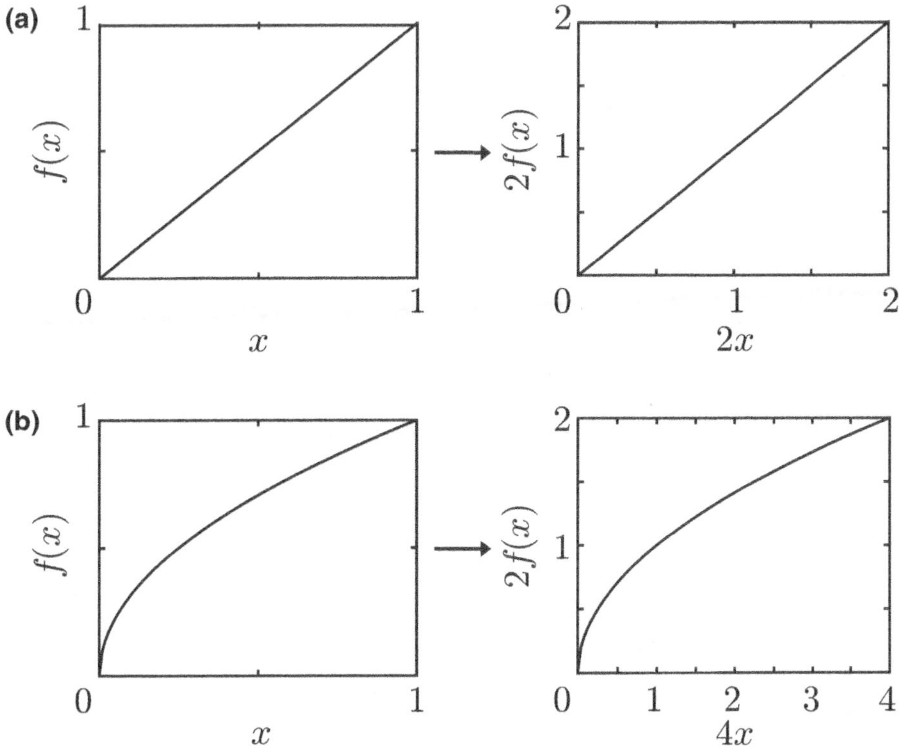

Fig. 2.3. Functions that exhibit (**a**) self-similar scaling behavior, and (**b**) self-affine scaling behavior. In (**a**), rescaling the axes of the graph by the same factor yields an identical curve. In (**b**), the axes must be rescaled by different factors to obtain an identical curve.

This function is a model for the height–height correlation function for a self-affine surface presented in Chap. 3, along with the definition of α, which, for the purpose of this discussion, is a constant. This function exhibits time-dependent scaling because we can choose the scale factors $s_1(t) = t^{-2\beta}$ and $s_2(t) = t^{\beta/\alpha}$ to give

$$t^{-2\beta} H\left(rt^{\beta/\alpha}, t\right) = 2\left[1 - \exp\left(-r^{2\alpha}\right)\right]. \tag{2.33}$$

This behavior is depicted in Fig. 2.4.

Another example of time-dependent scaling is in terms of the power spectral density function. The PSD of a self-affine surface can be modeled to have the form

$$P(k,t) = \frac{t^{2(\beta+1/z)}}{\left(1 + k^2 t^{2/z}\right)^{1+\alpha}}.$$

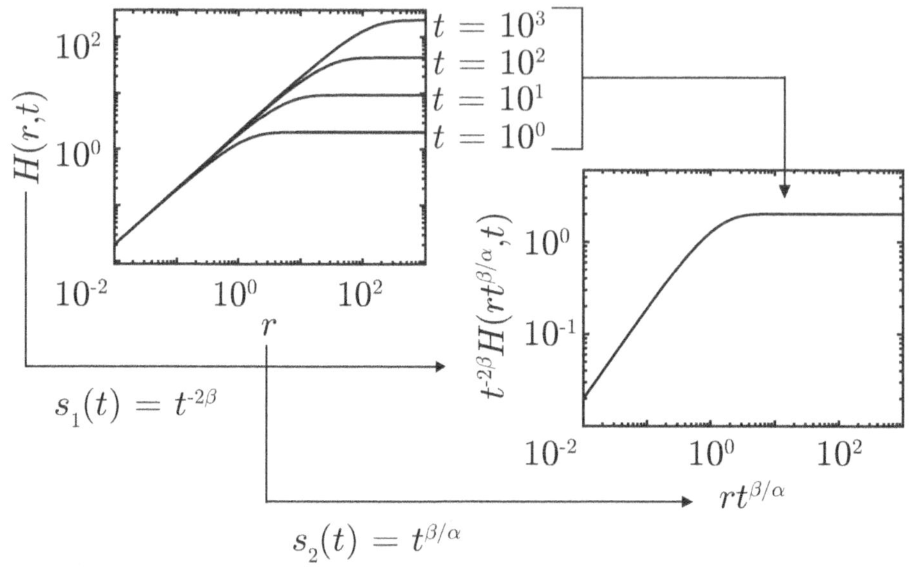

Fig. 2.4. Illustration of the time-dependent scaling of the function $H(r,t)$ given in (2.32). Scaling the horizontal axis of the graph by a factor of $t^{\beta/\alpha}$ and the vertical axis by a factor of $t^{-2\beta}$ results in a collapse of each curve onto one time-independent curve.

This PSD exhibits time-dependent scaling with scale factors $s_1(t) = t^{-2(\beta+1/z)}$ and $s_2(t) = t^{-1/z}$. However, the PSD of a mounded surface can be modeled with the form

$$P(k,t) = \frac{t^{2(\beta+1/z)}\left(1 + t^{2(1/z-p)} + k^2t^{2/z}\right)}{\left[\left(1 + t^{2(1/z-p)} + k^2t^{2/z}\right)^2 - \left(kt^{2/z-p}\right)^2\right]^{3/2}}.$$

An explicit time-dependence appears throughout this equation. If we choose the scale factors $s_1(t) = t^{-2(\beta+1/z)}$ and $s_2(t) = t^{-1/z}$, we obtain

$$t^{-2(\beta+1/z)}P\left(kt^{-1/z},t\right) = \frac{1 + t^{2(1/z-p)} + k^2}{\left[\left(1 + t^{2(1/z-p)} + k^2\right)^2 - \left(kt^{1/z-p}\right)^2\right]^{3/2}}.$$

This equation still depends explicitly on time, and there is no choice of scale factors that render this PSD time-independent in general. However, in the case of $p = 1/z$, the time-dependent factors in the scaled equation drop out, and we obtain

$$t^{-2(\beta+1/z)}P\left(kt^{-1/z},t\right) = \frac{1 + k^2}{\left[\left(1 + k^2\right)^2 - k^2\right]^{3/2}}, \quad \text{for } p = 1/z.$$

It is argued that the time-dependent scaling of surface correlation functions such as the height–height correlation function and power spectral density function are consequences of the dynamic scaling hypothesis. The time-dependent scaling behavior of a mounded PSD then implies that dynamic scaling does not hold unless $p = 1/z$.

2.9 Statistics from a Discrete Surface

Surface profiles obtained from numerical simulations or experimental measurement techniques often express the surface profile in a discrete form, where the domain of the surface is discretized into a certain number of lattice points at which the surface height is recorded. Therefore, it is useful to express the results from previous sections in a form that can be directly measured from a discrete surface profile. For the discussion that follows, consider a substrate of linear dimension L that is discretized uniformly into N lattice points per dimension, which implies that the continuous variable x can be expressed as the list of points

$$x \longrightarrow \left\{ x_i = i\frac{L}{N-1} \middle| \, i = 0, \ldots, N-1 \right\}.$$

Using this notation, the average $\langle f(\mathbf{x}) \rangle$ for a 1+1 dimensional surface can be expressed as the sum

$$\langle f(\mathbf{x}) \rangle \approx \frac{1}{N} \sum_{i=0}^{N-1} f(x_i), \tag{2.34}$$

and in 2+1 dimensions as

$$\langle f(\mathbf{x}) \rangle \approx \frac{1}{N^2} \sum_{i=0}^{N-1} \sum_{j=0}^{N-1} f(x_i, y_j). \tag{2.35}$$

For example, the autocorrelation function $R(r)$ can be computed discretely in 1+1 dimensions as

$$R(r) \approx \frac{1}{w^2 N} \sum_{i=0}^{N-1} h(x_i) h(x_i + r), \tag{2.36}$$

and in 2+1 dimensions as

$$R(r) \approx \frac{1}{w^2 N^2} \left\langle \sum_{i=0}^{N-1} \sum_{j=0}^{N-1} h(x_i, y_j) h(x_i + r_x, y_j + r_y) \right\rangle_{r=\sqrt{r_x^2 + r_y^2}}, \tag{2.37}$$

where the notation $\langle \cdots \rangle_{r=\sqrt{r_x^2+r_y^2}}$ means that the value of the double sum in (2.37) is averaged over all values of r_x and r_y that satisfy $r = \sqrt{r_x^2 + r_y^2}$. For instance, to compute $R(1)$, we must find the value of the double sum independently for $(r_x, r_y) = (1,0), (-1,0), (0,1)$, and $(0,-1)$, and then average the results together. Also, this form assumes periodic boundary conditions for the surface height, as the term $h(x_i + r_x, y_j + r_y)$ may exceed the boundaries of the lattice for large (r_x, r_y). To avoid using this condition, the autocorrelation function can only be measured from a subset of the original lattice by restricting the limits on the sums.

One correlation function that must be handled carefully for a discrete surface is the power spectral density function. The PSD of a discrete surface can be computed using a discrete Fourier transform, although algorithms exist that can compute this Fourier transform more efficiently, called fast Fourier transforms (FFT), which are implemented in commercially available software packages such as MATLAB. In addition, the PSD can be computed directly from the discrete version of the autocorrelation function, as the PSD is a Fourier transform of the autocorrelation function. When measuring the power spectral density function from a discrete surface, a few issues arise that are worth noting. First, if the surface has a nonzero mean height, a delta function behavior is introduced at $k = 0$, as can be seen from the definition of the PSD in (2.15),

$$\int \left(h(\mathbf{x},t) - \overline{h}\right) \mathrm{e}^{-\mathrm{i}\mathbf{k}\cdot\mathbf{x}} \mathrm{d}\mathbf{x} = \int h(\mathbf{x},t) \mathrm{e}^{-\mathrm{i}\mathbf{k}\cdot\mathbf{x}} \mathrm{d}\mathbf{x} - \overline{h}(2\pi)^d \delta^d(\mathbf{k}).$$

However, for a discrete surface, this delta function will not be measurable because the discrete lattice spacing a and lattice size L limit the range of measurable wavenumbers. Measuring wavenumbers above $k \sim a^{-1}$ and wavenumbers below $k \sim L^{-1}$ will not give meaningful results because the discrete nature of the lattice does not provide enough information to measure frequencies outside this range. Thus, the delta function behavior introduced by a nonzero mean will be outside the measurable frequency range, and will not affect the result of the measurement of the PSD.

In measuring surface statistics from a discrete surface of finite linear size L, a question arises as to the reliability of statistics measured from a finite-sized surface. If we consider the discrete surface as a sample from a surface that is infinitely large, clearly, as $L \to \infty$, the statistics measured from a discrete profile will converge to the true statistics of the surface. Thus, if we wish to draw meaningful conclusions from discrete statistics, we must choose L large enough to avoid sampling errors, but small enough to keep the amount of data manageable. To determine adequate bounds on L, we present an argument given in Yang et al. [174]. Consider the calculation of the mean height $\overline{h}$ from a discrete surface of linear size L,

$$\overline{h}_L = \frac{1}{L^d} \int_{-L/2}^{L/2} h(\mathbf{x}) \mathrm{d}\mathbf{x}, \tag{2.38}$$

where the limits of integration are the same in every dimension, and the origin of $\mathbf{x}$ has been chosen to be at the center of the discrete surface. The discrete sum that would appear in this expression for a discrete surface is approximated by an integral for simplicity. Clearly, the "true" mean height of the surface is the limit of this expression as $L \to \infty$. The limits of integration can be approximated by an exponential cutoff in the integration,

$$\frac{1}{L^d}\int_{-L/2}^{L/2} h(\mathbf{x})\mathrm{d}\mathbf{x} \approx \frac{1}{L^d}\int_{-\infty}^{\infty} h(\mathbf{x})\exp\left[-\frac{4|\mathbf{x}|^2}{L^2}\right]\mathrm{d}\mathbf{x}.$$

The uncertainty $\Delta\overline{h}_L$, which represents the standard deviation of the various values of $\overline{h}_L$ measured from a large number of different discrete surface profiles, is given by

$$\begin{aligned}(\Delta\overline{h}_L)^2 &= \left\langle \left[\overline{h}_L - \langle\overline{h}_L\rangle\right]^2 \right\rangle \\ &= \left\langle (\overline{h}_L)^2 - (\langle\overline{h}_L\rangle)^2 \right\rangle \\ &= \left\langle (\overline{h}_L)^2 \right\rangle,\end{aligned}$$

where the notation $\langle\cdots\rangle$ represents an ensemble average over many realizations of the discrete surface, and the last step follows from choosing the "true" mean height of the surface equal to zero. It follows that

$$(\Delta\overline{h}_L)^2 \approx \frac{1}{L^{2d}}\int\int\langle h(\mathbf{x})h(\mathbf{x}')\rangle\exp\left[-\frac{4|\mathbf{x}|^2}{L^2}\right]\exp\left[-\frac{4|\mathbf{x}'|^2}{L^2}\right]\mathrm{d}\mathbf{x}\mathrm{d}\mathbf{x}'.$$

The quantity $\langle h(\mathbf{x})h(\mathbf{x}')\rangle$ is related to the autocorrelation function for the surface. We can choose the model

$$\langle h(\mathbf{x})h(\mathbf{x}')\rangle = w^2\exp\left[-\frac{|\mathbf{x}-\mathbf{x}'|^2}{\xi^2}\right]$$

for the autocorrelation function, which is shown in Sect. 3.2 to be a model autocorrelation function for a self-affine surface with roughness exponent $\alpha = 1$. The uncertainty then becomes

$$(\Delta\overline{h}_L)^2 \approx \frac{w^2}{L^{2d}}\int\int\exp\left[-\frac{|\mathbf{x}-\mathbf{x}'|^2}{\xi^2}\right]\exp\left[-\frac{4|\mathbf{x}|^2}{L^2}\right]\exp\left[-\frac{4|\mathbf{x}'|^2}{L^2}\right]\mathrm{d}\mathbf{x}\mathrm{d}\mathbf{x}'.$$

Because the argument of this integral is a product of exponentials, we can evaluate the integral in one dimension, and raise the result to the power d, the dimension of the vectors $\mathbf{x}$ and $\mathbf{x}'$. Thus, with $a = \xi^{-2} + 4L^{-2}$,

$$(\Delta\overline{h}_L)^{2/d} \approx \frac{w^{2/d}}{L^2}\int\int\exp\left[-ax^2 - a(x')^2 + \frac{2xx'}{\xi^2}\right]\mathrm{d}x\mathrm{d}x'.$$

If we write the integral in x in terms of a perfect square,

$$\left(\Delta\bar{h}_L\right)^{2/d} \approx \frac{w^{2/d}}{L^2} \int \mathrm{d}x' \exp\left[-a\left(1-a^{-2}\xi^{-4}\right)(x')^2\right] \\ \times \int \mathrm{d}x \exp\left[-a\left(x - \frac{x'}{a\xi^2}\right)^2\right].$$

Using the result $\int \exp(-ax^2)\mathrm{d}x = \sqrt{\pi/a}$, the integral over x does not depend on x' because the limits of integration are infinite, which gives

$$\begin{aligned}\left(\Delta\bar{h}_L\right)^{2/d} &\approx \frac{w^{2/d}}{L^2}\sqrt{\frac{\pi}{a}} \int \mathrm{d}x' \exp\left[-a\left(1-a^{-2}\xi^{-4}\right)(x')^2\right] \\ &\approx \frac{w^{2/d}}{L^2}\frac{\pi}{\sqrt{a^2(1-a^{-2}\xi^{-4})}} \\ &\approx w^{2/d}\frac{\pi\xi}{4\sqrt{L^2/2+\xi^2}}.\end{aligned}$$

Therefore, the uncertainty in the discretely measured mean height is

$$\Delta\bar{h}_L \sim \frac{w\xi^{d/2}}{\left(L^2/2+\xi^2\right)^{d/4}}.$$

If $L \gg \xi$, this becomes

$$\Delta\bar{h}_L \sim w\sqrt{\frac{\xi^d}{L^d}}. \tag{2.39}$$

Thus, for the statistics obtained from a discrete surface profile to be representative of the actual surface statistics, we should have $\sqrt{\xi^d/L^d} \ll 1$. Because the lateral correlation length ξ is a natural length scale for the surface, it makes sense that we must average over a discrete surface that spans many correlation lengths to obtain reasonable statistics.

3

Self-Affine Surfaces

The study of self-affine surfaces forms the basis for the continuum study of all thin film rough surfaces. It is in the context of self-affine surfaces where the description of the surface is simplest and most elegant, and the ideas used to describe self-affine surfaces can be generalized to more complex surfaces such as mounded surfaces.

3.1 General Characteristics

Consider a surface profile $h(x)$ that, for r much less than the correlation length ξ, behaves as

$$|h(x+r) - h(x)| \sim (mr)^{\alpha}. \tag{3.1}$$

The term on the left-hand side of this equation represents the local roughness of the surface, and the exponent α is called the roughness exponent for the surface. The local slope of the surface profile is denoted by m. In higher dimensions, for an isotropic surface, this relation becomes

$$|h(\mathbf{x}+\mathbf{r}) - h(\mathbf{x})| \sim (m|\mathbf{r}|)^{\alpha}. \tag{3.2}$$

In one dimension, it follows that the relation

$$|h(\varepsilon x + \varepsilon r) - h(\varepsilon x)| \sim (\varepsilon m r)^{\alpha},$$

also holds, which can be rearranged to give

$$|\varepsilon^{-\alpha} h(\varepsilon x + \varepsilon r) - \varepsilon^{-\alpha} h(\varepsilon x)| \sim (mr)^{\alpha}.$$

By comparison with (3.1), this implies that the height profile can be expressed as

$$h(x) \sim \varepsilon^{-\alpha} h(\varepsilon x). \tag{3.3}$$

Such a surface profile is said to be self-affine [100], and the roughness exponent α characterizes the short-range roughness of a self-affine surface, with larger

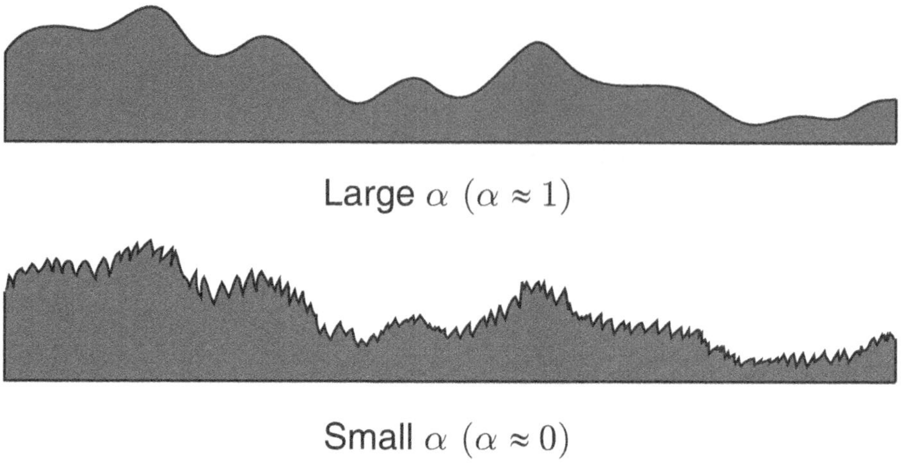

Fig. 3.1. This diagram shows a comparison of the local surface morphology for surfaces with similar values for the interface width w, but different values of α. A smaller value of α implies a rougher local surface, where α lies between 0 and 1.

values of α representing a smoother local surface profile [20]. Surfaces with different values of α are depicted in Fig. 3.1.

It is noted that α lies in the range $0 \leq \alpha \leq 1$. To derive this condition, consider two length scales, x and $x' = \varepsilon x$. The surface slope on each of these length scales is approximately given by $\partial h/\partial x$ and $\partial h/\partial x'$, respectively from (2.11). However, from the definition of a self-affine surface,

$$\frac{\partial h}{\partial x} \sim \frac{\partial}{\partial x}\left(\varepsilon^{-\alpha} h(\varepsilon x)\right) = \varepsilon^{-\alpha} \frac{\partial h(\varepsilon x)}{\partial x} = \varepsilon^{1-\alpha} \frac{\partial h}{\partial x'}. \tag{3.4}$$

If $\varepsilon \geq 1$, the x' length scale is more "stretched out" than the x length scale, which implies that the surface slope on the x' length scale is smaller than the surface slope on the x length scale. To satisfy this requirement, from (3.4), $\varepsilon^{1-\alpha} \geq 1$ for $\varepsilon \geq 1$, which gives $1 - \alpha \geq 0$, or $\alpha \leq 1$. In addition, for (3.1) to be physical in the limit as $r \to 0$, $\lim_{r \to 0} r^{\alpha} = 0$, which gives $\alpha \geq 0$. In the specific case where $\alpha = 1$, the surface is said to exhibit self-similar scaling because the scale factors in the horizontal and vertical directions are equal. This scaling behavior is reminiscent of the definition of a fractal.

It is important to mention that a real thin film surface will only exhibit self-affine behavior over a certain range of length scales, and there exists a cutoff length scale, a, beneath which the surface may not be self-affine. For example, once the length scale becomes smaller than the size of an atom, the surface height is no longer well defined, and the surface cannot be self-affine. For the discussions that follow, if the cutoff length a is much smaller than the correlation length ξ of the surface, we can treat the surface as if $a \to 0$, and assume self-affinity on all scales for simplicity.

To make the connection between a self-affine surface and a fractal, we present a general introduction of fractal behavior. We can define a fractal for our purposes as follows. If we are interested in finding the area of an object in two dimensions, we can cover the object with small patches of linear size l and count how many patches it takes to fully cover the object. If we found that it takes N patches to cover the object, the area of the object would be the number of patches used times the area of each patch, which would be

$$A = Nl^2. \tag{3.5}$$

We can do the same thing in one and three dimensions, and denoting this embedding Euclidean dimension by d, the "area" would satisfy

$$A = Nl^d, \tag{3.6}$$

where "area" in one dimension is length, and "area" in three dimensions is volume. Because we are dealing mainly in two dimensions, we continue with the concept of area. If we repeat this procedure with different size patches by changing l, we will find that the number of patches we use to cover the surface is related to the size of the patch as

$$N(l) \sim l^{-D}, \tag{3.7}$$

where D is the fractal dimension of the surface. For ordinary surfaces, the fractal dimension D is equal to the embedding Euclidean dimension d because the area does not depend on the size of the coverings used to measure it. For example, in two dimensions, a square of side length L can be covered with smaller squares of side length $l < L$. The number of smaller squares $N(l)$ required to cover the large square satisfies $N(l)l^2 = L^2$, which implies that $N(l) \sim l^{-2}$, and the Euclidean dimension is equal to the fractal dimension. A fractal is a surface where the surface area one measures depends on the size l of patches used to measure it. The area of the square in the previous example did not change when the patch size changed; it remained constant at L^2. However, there are surfaces where the area measured depends on the length scale used to measure it, and self-affine surfaces belong to this class of surfaces.

To show that self-affine surfaces behave as fractals, we can write the surface area of a thin film described by the height profile $h(\mathbf{x})$ as

$$A = \int \sqrt{1 + |\nabla h(\mathbf{x})|^2} \mathrm{d}\mathbf{x}. \tag{3.8}$$

However, for a self-affine surface that obeys (3.1), if l is the size of the patch being used to measure the surface area, then

$$|\nabla h(x)| \approx \frac{h(x+l) - h(x)}{l} \sim \frac{l^\alpha}{l} \sim l^{\alpha - 1}. \tag{3.9}$$

Because $0 \leq \alpha \leq 1$, if l is small, then $l^{\alpha-1}$ will be very large, and the argument of the integral for the surface area can be approximated as

$$A \approx \int |\nabla h(\mathbf{x})| \mathrm{d}\mathbf{x} \sim \int l^{\alpha-1} \mathrm{d}\mathbf{x} \sim l^{\alpha-1}, \tag{3.10}$$

where the integral $\int \mathrm{d}\mathbf{x}$ does not depend on l. From (3.6) and (3.7), this gives

$$A \sim l^{\alpha-1} \sim l^{-D} l^{d}, \tag{3.11}$$

which implies that $D = d + 1 - \alpha$. Therefore, for a surface with $\alpha = 1$, the fractal dimension and embedding Euclidean dimension are equal. However, if $\alpha < 1$, the surface area measured depends on the size of the patch used to measure it. From (3.10), the area depends on the size l of the patch as $A \sim l^{\alpha-1}$, so the smaller the size of the patch, the larger the measured area! The most commonly cited real-world example of this phenomenon is the measurement of the length of a coastline. If you measure the length of the east coast of the United States with a ruler on a map, you will measure a much smaller distance than if you walked the east coast with a ruler, measuring the coastline along the way. The smaller the device, or "patch", you use to measure the length of the coastline, the larger the length you will measure because larger measuring devices miss the detailed structure of the coastline that finer measuring devices catch. This behavior is why the definition of the surface slope in Sect. 2.6 diverges when $\alpha \neq 1$. The derivative of the height profile is not well defined in a continuum sense because the value of the derivative depends on the length scale used to measure it. In other words, the limit

$$\lim_{l \to 0} \frac{h(x+l) - h(x)}{l}, \tag{3.12}$$

that is used to define the derivative, behaves as $l^{\alpha-1}$, which becomes infinite if $\alpha < 1$. As previously discussed, there is a cutoff length scale a beneath which the surface is no longer self-affine, so the divergence of the derivative is simply a result of discussing surface statistics in the limit as $a \to 0$, which does not hold for realistic surfaces. However, when applying continuum statistics such as the autocorrelation function and height–height correlation function to self-affine surfaces, the continuum approximation can be used to obtain a value for the local slope m, and the definition of the local slope m in (3.1) is discussed in Sect. 3.3.

3.2 Lateral Correlation Functions

When the surface height profile obeys (3.3), the correlation functions have similar scaling properties. For small r, substituting (3.1) into the definition of the height–height correlation function from (2.9) yields

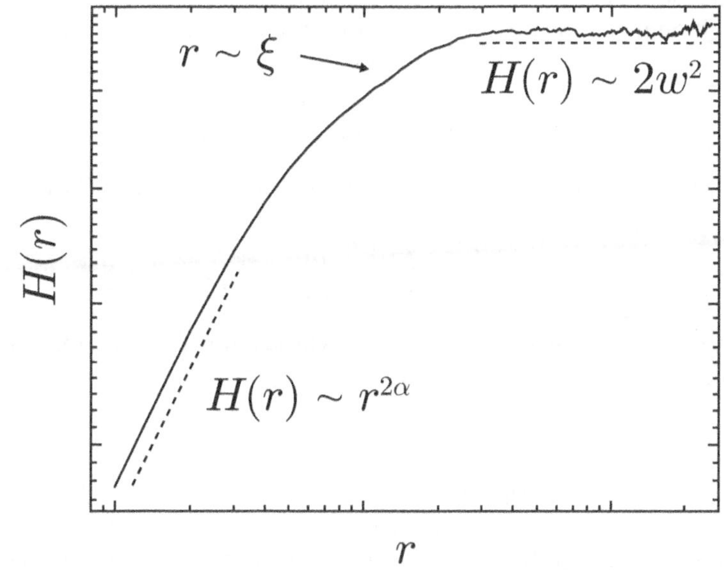

Fig. 3.2. Representative height–height correlation function obtained from a simulated self-affine surface. The plot is on a log–log scale, which gives the height–height correlation function a linear behavior for small r with slope 2α.

$$H(r) = \left\langle |h(x+r) - h(x)|^2 \right\rangle \sim \left\langle |(mr)^\alpha|^2 \right\rangle \sim (mr)^{2\alpha}. \tag{3.13}$$

It follows that the height–height correlation function behaves as

$$H(r) \propto \begin{cases} (mr)^{2\alpha}, & r \ll \xi, \\ 2w^2, & r \gg \xi. \end{cases} \tag{3.14}$$

For a self-affine surface, the height–height correlation function $H(r)$ can be expressed in the scaling form

$$H(r) = 2w^2 f\left(\frac{r}{\xi}\right), \tag{3.15}$$

where the function f behaves as

$$f(x) = \begin{cases} x^{2\alpha}, & x \ll 1, \\ 1, & x \gg 1. \end{cases} \tag{3.16}$$

This behavior is seen in Fig. 3.2, which depicts a representative height–height correlation function for a self-affine surface. Note that the height–height correlation function for small r behaves as a power law.

Several analytic forms for the height–height correlation have been proposed that satisfy the requirements for a self-affine surface given in (3.14). For an isotropic self-affine surface, Sinha et al. [142] proposed the functional form

$$H(r) = 2w^2 \left[1 - \exp\left[-\left(\frac{r}{\xi}\right)^{2\alpha}\right]\right]. \tag{3.17}$$

This form satisfies (3.14) because an expansion of the exponential for $r \ll \xi$ gives

$$H(r) \approx 2w^2 \left[1 - \left(1 - \left(\frac{r}{\xi}\right)^{2\alpha}\right)\right] \approx \frac{2w^2}{\xi^{2\alpha}} r^{2\alpha} \sim r^{2\alpha}.$$

From (2.10), this implies that the autocorrelation function $R(r)$ can be expressed as

$$R(r) = \exp\left[-\left(\frac{r}{\xi}\right)^{2\alpha}\right]. \tag{3.18}$$

We refer to this model as the exponential correlation model. Unfortunately, using the exponential correlation model for $R(r)$ does not work as $\alpha \to 0$ because the autocorrelation function becomes constant when $\alpha = 0$, which does not reflect the required behavior of the function for a self-affine surface. To remedy this, a more complicated autocorrelation function has been proposed [118] called the K-correlation model,

$$R(r) = \frac{\alpha}{2^{\alpha-1}\Gamma(\alpha+1)} \left(\frac{r}{\xi}\sqrt{2\alpha}\right)^{\alpha} K_\alpha\left(\frac{r}{\xi}\sqrt{2\alpha}\right), \tag{3.19}$$

where $\Gamma(x)$ is the gamma function and $K_\alpha(x)$ is the α-order modified Bessel function of the second kind. The gamma function and modified Bessel function of the second kind are discussed in App. A.

Let us verify that the K-correlation model satisfies the properties required of the height–height correlation function of a self-affine surface given in (3.14). From Sect. A.1.3, for $r \ll \xi$ and $0 < \alpha < 1$, the modified Bessel function of the second kind behaves as

$$K_\alpha\left(\frac{r}{\xi}\sqrt{2\alpha}\right) \sim \frac{\Gamma(\alpha)}{2}\left(\frac{r}{2\xi}\sqrt{2\alpha}\right)^{-\alpha} - \frac{\Gamma(1-\alpha)}{2\alpha}\left(\frac{r}{2\xi}\sqrt{2\alpha}\right)^{\alpha}.$$

Using this result, the autocorrelation function behaves as

$$R(r) \approx 1 - \left(\frac{\alpha}{2}\right)^{\alpha} \frac{\Gamma(1-\alpha)}{\Gamma(1+\alpha)} \left(\frac{r}{\xi}\right)^{2\alpha}.$$

It follows that the height–height correlation function behaves as

$$H(r) \approx 2w^2 \left[1 - \left(1 - \left(\frac{\alpha}{2}\right)^{\alpha} \frac{\Gamma(1-\alpha)}{\Gamma(1+\alpha)} \left(\frac{r}{\xi}\right)^{2\alpha}\right)\right] \sim r^{2\alpha}.$$

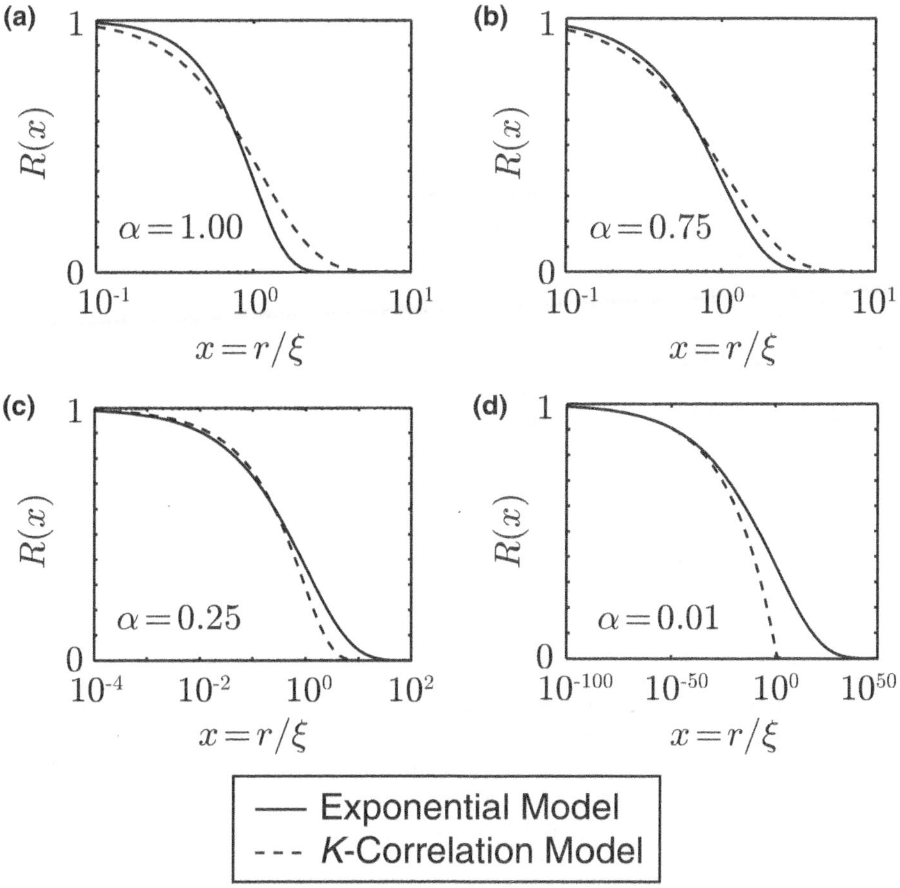

Fig. 3.3. Comparison of two proposed forms of the autocorrelation function for a self-affine surface given in (3.18), the exponential model, and (3.19), the K-correlation model. Each plot has a different value of α, (**a**) $\alpha = 1.00$, (**b**) $\alpha = 0.75$, (**c**) $\alpha = 0.25$, and (**d**) $\alpha = 0.01$.

In addition, because $K_\alpha(x)$ has an exponentially decaying behavior for large x, $H(r) \approx 2w^2$ for $r \gg \xi$. Therefore, (3.19) is a valid autocorrelation function for a self-affine surface. The advantage of this form of the autocorrelation function is that it possesses an analytic Fourier transform, and thus its PSD can be expressed in terms of elementary functions.

A comparison of the exponential correlation model given in (3.18) and the K-correlation model in (3.19) is pictured in Fig. 3.3 for various values of α. For $\alpha > \frac{1}{2}$, the K-correlation model approaches zero more gradually than does the exponential model, and for $\alpha < \frac{1}{2}$, the K-correlation model approaches zero more abruptly than the exponential model. A crossover occurs at $\alpha = \frac{1}{2}$ because the two models are equal when $\alpha = \frac{1}{2}$, which follows from the representation of the modified Bessel function of the second kind

$$K_{1/2}(x) = \sqrt{\frac{\pi}{2x}} \mathrm{e}^{-x},$$

and the value of the gamma function, $\Gamma\left(\frac{3}{2}\right) = \sqrt{\pi}/2$. Keep in mind that the exponential model and K-correlation model are only models for the correlation functions, and any height–height correlation function that satisfies (3.14) may be considered as a model [119, 120].

3.3 Local Slope

In the previous section, it was shown that the small r behavior of the height–height correlation function is

$$H(r) \sim (mr)^{2\alpha}, \quad \text{for } r \ll \xi, \tag{3.20}$$

which depends on the local slope m. Motivated by this behavior, we can define the local slope as

$$m^{2\alpha} \equiv \left[r^{-2\alpha} H(\mathbf{r})\right]_{\mathbf{r}=\mathbf{0}} = 2w^2 \left[r^{-2\alpha}\left(1 - R(\mathbf{r})\right)\right]_{\mathbf{r}=\mathbf{0}}, \tag{3.21}$$

where (2.10) was used to relate the height–height correlation function to the autocorrelation function. By dimensional analysis, we can deduce from this expression that the local slope behaves as

$$m \sim \frac{w^{1/\alpha}}{\xi}, \tag{3.22}$$

because the height–height correlation function has units of w^2, and is multiplied by a distance to the power -2α. Using (3.21), if the autocorrelation function $R(r) \approx 1 - cr^{2\alpha}$ for small r, the local slope m is given by

$$m = (2w^2 c)^{1/(2\alpha)}. \tag{3.23}$$

With this result, the exponential model from (3.18) gives a local slope of

$$m = \frac{(w\sqrt{2})^{1/\alpha}}{\xi}, \tag{3.24}$$

whereas the K-correlation model in (3.19) gives a local slope of

$$m = \frac{(w\sqrt{2})^{1/\alpha}}{\xi} \cdot \sqrt{\frac{\alpha}{2}} \left[\frac{\Gamma(1-\alpha)}{\Gamma(1+\alpha)}\right]^{1/(2\alpha)}. \tag{3.25}$$

Note that the local slope in both cases behaves as (3.22).

In Sect. 2.6, the RMS surface slope ϱ of a surface was given as

$$\varrho^2 = \left\langle |\nabla h(\mathbf{x})|^2 \right\rangle . \tag{3.26}$$

From the definition of a self-affine surface given in (3.1), this equation can be expressed as

$$\begin{aligned}\varrho^2 &\sim \lim_{r\to 0} \left\langle \left| \frac{h(x+r)-h(x)}{r} \right|^2 \right\rangle \\ &\sim \lim_{r\to 0} \left\langle \frac{|(mr)^\alpha|^2}{r^2} \right\rangle \\ &\sim \lim_{r\to 0} m^{2\alpha} r^{2\alpha-2} .\end{aligned} \tag{3.27}$$

If $\alpha < 1$, the definition of the slope as ϱ in (3.26) will diverge. From the discussion of the behavior of ϱ in Sect. 2.6, it follows that

$$m^{2\alpha} \sim \lim_{r\to 0} r^{2-2\alpha} \varrho^2 \sim -w^2 \left[r^{2-2\alpha} \nabla_{\mathbf{r}}^2 R(\mathbf{r}) \right]_{\mathbf{r}=\mathbf{0}} . \tag{3.28}$$

This expression relates the local slope m to the surface slope ϱ. One can also take this expression as a definition of the local slope m, which may differ by a constant factor from (3.21). If the surface is isotropic in 2+1 dimensions, this expression becomes

$$m^{2\alpha} \sim -w^2 \left[r^{2-2\alpha} \left(\frac{\mathrm{d}^2}{\mathrm{d}r^2} + \frac{1}{r}\frac{\mathrm{d}}{\mathrm{d}r} \right) R(r) \right]_{r=0} . \tag{3.29}$$

This expression is consistent with (3.21) in 2+1 dimensions by adding a factor of $2\alpha^2$,

$$m^{2\alpha} = -\frac{w^2}{2\alpha^2} \left[r^{2-2\alpha} \left(\frac{\mathrm{d}^2}{\mathrm{d}r^2} + \frac{1}{r}\frac{\mathrm{d}}{\mathrm{d}r} \right) R(r) \right]_{r=0} , \tag{3.30}$$

which can be shown by substituting the form $R(r) \approx 1 - cr^{2\alpha}$, and comparing to (3.23).

3.4 Power Spectral Density Function

The PSD of a self-affine surface can be quantified by its asymptotic behavior, as was discussed for the height–height correlation function in Sect. 3.2. The PSD can be expressed as

$$P(k) = \frac{w^2}{(2\pi)^d} \int R(r) \mathrm{e}^{\mathrm{i}\mathbf{k}\cdot\mathbf{r}} \mathrm{d}\mathbf{r}, \tag{3.31}$$

where $k = |\mathbf{k}|$ and $r = |\mathbf{r}|$ for an isotropic surface. Using one of Green's identities over the closed volume Σ, which is a generalization of integration by parts in arbitrary dimensions,

$$\int_{\Sigma} \psi(\mathbf{r})\nabla^2\phi(\mathbf{r})\mathrm{d}\mathbf{r} = \int_{\partial\Sigma} \psi(\mathbf{r})\frac{\partial\phi(\mathbf{r})}{\partial n}\mathrm{d}S - \int_{\Sigma} \nabla\psi(\mathbf{r})\cdot\nabla\phi(\mathbf{r})\mathrm{d}\mathbf{r}.$$

If we take $\psi(\mathbf{r}) = R(r)$ and $\phi(\mathbf{r}) = -k^{-2}\mathrm{e}^{\mathrm{i}\mathbf{k}\cdot\mathbf{r}}$, this expression becomes

$$\int_{\Sigma} R(r)\mathrm{e}^{\mathrm{i}\mathbf{k}\cdot\mathbf{r}}\mathrm{d}\mathbf{r} = \int_{\partial\Sigma} R(r)\frac{\partial}{\partial n}\left[-\frac{1}{k^2}\mathrm{e}^{\mathrm{i}\mathbf{k}\cdot\mathbf{r}}\right]\mathrm{d}S + \frac{1}{k^2}\int_{\Sigma}\nabla R(r)\cdot\left(\mathrm{i}\mathbf{k}\mathrm{e}^{\mathrm{i}\mathbf{k}\cdot\mathbf{r}}\right)\mathrm{d}\mathbf{r}.$$

In the limit of an infinite domain, $R(r) \to 0$ and the surface integral vanishes, which gives for the behavior of the PSD,

$$P(k) \sim w^2\frac{\hat{\mathbf{k}}}{k}\cdot\int \nabla R(r)\mathrm{e}^{\mathrm{i}\mathbf{k}\cdot\mathbf{r}}\mathrm{d}\mathbf{r}.$$

The presence of the autocorrelation function in this expression separates the integral into two regimes, $r \le \xi$ and $r > \xi$. Because $R(r)$ is significant only for $r \le \xi$, we can approximate this integral as

$$P(k) \sim w^2\frac{\hat{\mathbf{k}}}{k}\cdot\int_0^{\xi} \nabla R(r)\mathrm{e}^{\mathrm{i}\mathbf{k}\cdot\mathbf{r}}r^{d-1}\mathrm{d}r. \tag{3.32}$$

The volume element in d dimensions $\mathrm{d}\mathbf{r}$ is proportional to $r^{d-1}\mathrm{d}r$. In a rough approximation, for large $\mathbf{k}\cdot\mathbf{r}$, the exponential oscillates much faster than the rest of the integrand, which has the effect of averaging the integral out to zero for large $\mathbf{k}\cdot\mathbf{r}$. Thus, the integral is cut off when $\mathbf{k}\cdot\mathbf{r} \sim kr \approx 1$, and the integration is significant only over the domain $r \in [0, k^{-1}]$. In the regime where $k \gg \xi^{-1}$, the domain $[0, k^{-1}]$ cuts off the domain of integration in (3.32), which gives

$$P\left(k \gg \xi^{-1}\right) \sim w^2\frac{\hat{\mathbf{k}}}{k}\cdot\int_0^{k^{-1}} \nabla R(r)r^{d-1}\mathrm{d}r.$$

If we change to the dimensionless variable $x = r\xi^{-1}$, this integral becomes

$$P\left(k \gg \xi^{-1}\right) \sim w^2\xi^d\frac{\hat{\mathbf{k}}}{k\xi}\cdot\int_0^{(k\xi)^{-1}} \nabla R(x)x^{d-1}\mathrm{d}x.$$

To obtain this form, recall that the gradient also introduces a factor of ξ, $\nabla_r R(r) = \xi^{-1}\nabla_x R(x)$. For a self-affine surface, $\nabla R(x) \sim \nabla H(x) \approx x^{2\alpha-1}\hat{\mathbf{x}}$ when $x \ll 1$. The PSD then behaves as

$$\begin{aligned} P\left(k \gg \xi^{-1}\right) &\sim \frac{w^2\xi^d}{k\xi}\int_0^{(k\xi)^{-1}} x^{2\alpha+d-2}\mathrm{d}x \\ &\sim \frac{w^2\xi^d}{k\xi}\left[x^{2\alpha+d-1}\right]_0^{(k\xi)^{-1}} \\ &\sim w^2\xi^d(k\xi)^{-2\alpha-d}. \end{aligned} \tag{3.33}$$

For small k, $k \ll \xi^{-1}$, we can no longer use (3.32) because of the factor of k^{-1} in front of the integral and we go back to (3.31). The exponential restricts the domain to $r \in [0, k^{-1}]$, however, this is less restrictive than $r \in [0, \xi]$ for small k, which gives

$$P\left(k \ll \xi^{-1}\right) \sim w^2 \int_0^{\xi} R(r) r^{d-1} \mathrm{d}r \sim w^2 \xi^d. \tag{3.34}$$

This result is independent of k, and the behavior of ξ can be found through dimensional analysis; the autocorrelation function is dimensionless, and the integral has dimensions of length to the power d. We can summarize these results by expressing the PSD of a self-affine surface in a scaling form,

$$P(k) = w^2 \xi^d g(k\xi), \tag{3.35}$$

where

$$g(x) \propto \begin{cases} 1, & x \ll 1, \\ x^{-2\alpha-d}, & x \gg 1. \end{cases} \tag{3.36}$$

Using the form of the autocorrelation function given by the K-correlation model in (3.19), the PSD of a self-affine surface in 2+1 dimensions can be modeled as

$$P(k) = \frac{w^2 \xi^2}{2\pi \left(1 + \frac{k^2 \xi^2}{2\alpha}\right)^{1+\alpha}}. \tag{3.37}$$

The mathematics of calculating this form of the PSD are given in Sect. A.4.2. The asymptotic behavior of the PSD is given by, for $k \gg \xi^{-1}$,

$$P(k) \approx \frac{w^2 \xi^2}{2\pi \left(\frac{k^2 \xi^2}{2\alpha}\right)^{1+\alpha}} \propto k^{-2-2\alpha}. \tag{3.38}$$

This behavior is seen in Fig. 3.4. Note that the PSD has no characteristic peak, which allows for the scaling definition of a self-affine surface [187]. A characteristic peak in the PSD implies that there is a characteristic length scale on the surface that will change upon rescaling, breaking the scaling behavior of the surface. Because self-affine surfaces have no such peak in their PSD, the scaling definition holds.

3.5 Dynamic Scaling

A surface profile is said to exhibit dynamic scaling if the surface height profile can be scaled in time. For a self-affine surface, this gives [8, 40, 41, 104]

$$h(x,t) \sim \varepsilon^{-\alpha} h(\varepsilon x, \varepsilon^z t), \tag{3.39}$$

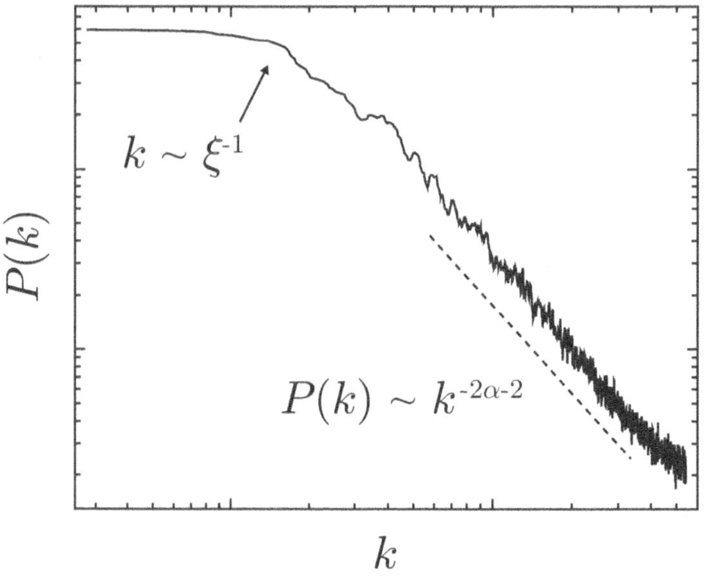

Fig. 3.4. Representative power spectral density function (PSD) obtained from a simulated self-affine surface in 2+1 dimensions ($d = 2$). The PSD spectrum exhibits no characteristic peak.

where z is the dynamic exponent. The 1+1 dimensional form of the height profile has been used for simplicity, the same concept can be extended to 2+1 dimensions. If this scaling in time holds, increasing the time by a factor ε increases the horizontal length scale by a factor $\varepsilon^{1/z}$. Thus, the lateral correlation length, which is a function of the horizontal correlations on the surface, must evolve as

$$\xi(t) \sim t^{1/z}. \tag{3.40}$$

Similarly, increasing the time by a factor ε changes the vertical length scale by a factor $\varepsilon^{\alpha/z}$. The interface width is a function of the vertical height profile of the surface, therefore the interface width must evolve as

$$w(t) \sim t^{\alpha/z}. \tag{3.41}$$

Dynamic scaling predicts an interesting behavior for the time evolution of the surface. All parameters that measure the surface are related to one another, because the surface profile scales as a whole in time. In particular, since the surface grows on a substrate of finite linear size L, there is a natural bound on the growth of the lateral correlation length because surface heights cannot be correlated beyond the size of the substrate L. This implies that there exists a crossover time t_x where the lateral correlation length saturates, given by

$$\xi(t_x) \sim t_x^{1/z} \sim L \Rightarrow t_x \sim L^z.$$

However, because the surface obeys dynamic scaling, if the lateral correlation length saturates, so must the interface width or else the scaling behavior of the surface will break down. Thus, the interface width must also saturate at the crossover time, which gives an expression for the saturation value of the interface width,

$$w_{sat} \sim t_x^{\alpha/z} \sim (L^z)^{\alpha/z} \sim L^\alpha.$$

In (2.4), the interface width was defined as evolving with an exponent β, which implies that the characteristic behavior of the interface width under dynamic scaling can be expressed as

$$w(t) \sim \begin{cases} t^\beta, & t \ll t_x \sim L^z, \\ L^\alpha, & t \gg t_x \sim L^z. \end{cases} \tag{3.42}$$

Comparing (3.42) with (3.41) gives the well-known relationship between the scaling exponents under dynamic scaling,

$$z = \frac{\alpha}{\beta}. \tag{3.43}$$

In defining dynamic scaling, one can begin with the hypothesis that the interface width behaves in this manner – growing as a power law before the crossover time and saturating afterwards – and reach the same conclusions presented here [8].

Experimentally, one can measure the values for α, β, and z from different surface statistics: α from the short-range behavior of the height–height correlation function, β from the time evolution of the interface width, and z from the time evolution of the lateral correlation length. These three exponents characterize the behavior of the surface and are related in a specific manner, essentially simplifying the problem of characterizing a self-affine surface to finding values for these exponents. However, the assumptions made when defining self-affinity and dynamic scaling do not hold in general, and for mounded surfaces more information is needed to fully describe the surface profile [123]. Nevertheless, a wide range of surfaces grown under various techniques obey self-affinity and dynamic scaling, for example, experimentally deposited surfaces grown under normal incidence thermal evaporation [175], and simulated surface profiles that grow according to local stochastic continuum equations discussed in Chap. 5.

3.5.1 Stationary and Nonstationary Growth

From (3.22), the local slope m of a self-affine surface behaves as

$$m \propto \frac{w^{1/\alpha}}{\xi} \sim t^{\beta/\alpha - 1/z}. \tag{3.44}$$

However, under dynamic scaling from (3.43), $\beta/\alpha - 1/z = 0$, and dynamic scaling predicts that the local slope does not change with time. Growth for which the local slope is constant is said to be stationary. A nonstationary local slope indicates that dynamic scaling does not rigorously hold for a surface, and certain investigations in self-affine surface growth have found a logarithmic behavior at large times for the local slope [91],

$$m(t) \sim \sqrt{\ln t}. \tag{3.45}$$

This growth is referred to as nonstationary growth because the local slope changes with time. However, a logarithmic evolution is very slow compared to a power law as

$$\lim_{t \to \infty} \frac{\sqrt{\ln t}}{t^{\delta}} = 0 \quad \text{for any } \delta > 0. \tag{3.46}$$

In fact, such a logarithmic behavior can loosely be considered to be a power law with a vanishingly small exponent, as

$$t^{\delta} = \exp[\delta \ln t] \approx 1 + \delta \ln t, \tag{3.47}$$

for $|\ln t| \ll \delta^{-1}$, which is a significant domain if δ is very small. Often, as is the case with the Edwards–Wilkinson model discussed in Sect. 5.1.2, a power-law exponent of zero can be interpreted as a logarithm. Thus, a logarithmic behavior for the local slope is essentially constant when compared to the power-law growth of other surface statistics because the logarithm grows so slowly. As the local slope evolves with exponent $\beta/\alpha - 1/z$, a logarithmic growth for the local slope can be approximated by setting this exponent equal to zero, as would be the prediction under stationary growth. Therefore, surface statistics for a surface evolving under nonstationary growth dynamics will fit with the predictions of dynamic scaling if the local slope evolves logarithmically in time. If the local slope exhibits a power-law behavior in time, the roughness evolution can be described in the context of anomalous scaling, described in Sect. 3.5.3.

3.5.2 Time-Dependent Scaling

Recall from (3.14) that the height–height correlation function for a self-affine surface behaves as

$$H(r) \sim \begin{cases} (mr)^{2\alpha}, & r\xi^{-1} \ll 1, \\ 2w^2, & r\xi^{-1} \gg 1. \end{cases}$$

If the surface also obeys dynamic scaling, we can write this expression as

$$H(r,t) \sim \begin{cases} (mr)^{2\alpha}, & rt^{-1/z} \ll 1, \\ 2t^{2\beta}, & rt^{-1/z} \gg 1. \end{cases}$$

Therefore, if we follow the discussion of Sect. 2.8.2 and define the time-dependent scale factors $s_1(t) = t^{-2\beta}$ and $s_2(t) = t^{1/z}$, the height–height correlation function becomes

$$t^{-2\beta} H\left(rt^{1/z}, t\right) \sim \begin{cases} (mr)^{2\alpha}, & r \ll 1, \\ 2, & r \gg 1. \end{cases} \tag{3.48}$$

This form follows because $\beta - \alpha/z = 0$ from dynamic scaling. Thus, the height–height correlation function exhibits time-dependent scaling when the surface obeys dynamic scaling and the local slope m is time-independent as is the case in stationary growth. This behavior is included in Fig. 2.4, where it was given as an example of time-dependent scaling.

A similar behavior can be observed for the power spectral density function. From (3.35), for a self-affine surface, the PSD behaves as

$$P(k) \sim \begin{cases} w^2 \xi^d, & k\xi \ll 1, \\ w^2 \xi^d (k\xi)^{-2\alpha-d}, & k\xi \gg 1. \end{cases}$$

Under dynamic scaling, the time dependence of the PSD is given by

$$P(k,t) \sim \begin{cases} t^{2\beta+d/z}, & kt^{1/z} \ll 1, \\ t^{2\beta+d/z}(kt^{1/z})^{-2\alpha-d}, & kt^{1/z} \gg 1. \end{cases} \tag{3.49}$$

We can clearly choose the scale factors $s_1(t) = t^{-2\beta-d/z}$ and $s_2(t) = t^{-1/z}$ to give the time-independent form

$$t^{-2\beta-d/z} P\left(kt^{-1/z}, t\right) \sim \begin{cases} 1, & k \ll 1, \\ k^{-2\alpha-d}, & k \gg 1. \end{cases} \tag{3.50}$$

This scaling is pictured in Fig. 3.5. The observation that surface correlation functions such as the height–height correlation function and power spectral density exhibit time-dependent scaling when the surface obeys dynamic scaling should come as no surprise. Dynamic scaling predicts that the statistical properties of a surface can be scaled in time. Surface correlation functions can be considered statistics themselves, and as such must scale in time under dynamic scaling.

3.5.3 Anomalous Scaling

It should be noted that another scaling hypothesis has also been proposed, called anomalous scaling [87, 90, 138], that builds on the scaling relations predicted by dynamic scaling. Anomalous scaling predicts that the global interface width w depends on both the system size L and time t as predicted by dynamic scaling in (3.42), but the local interface width depends both on the measurement window size $l < L$ and the time t as

$$w(l,t) \sim \begin{cases} t^{\beta}, & t \ll l^z, \\ t^{\kappa} l^{\alpha_{loc}}, & t \gg l^z, \end{cases} \tag{3.51}$$

where $\kappa = \beta - \alpha_{loc}/z$, and α_{loc} is a local roughness exponent that differs in general from α. This behavior for the interface width implies that there is

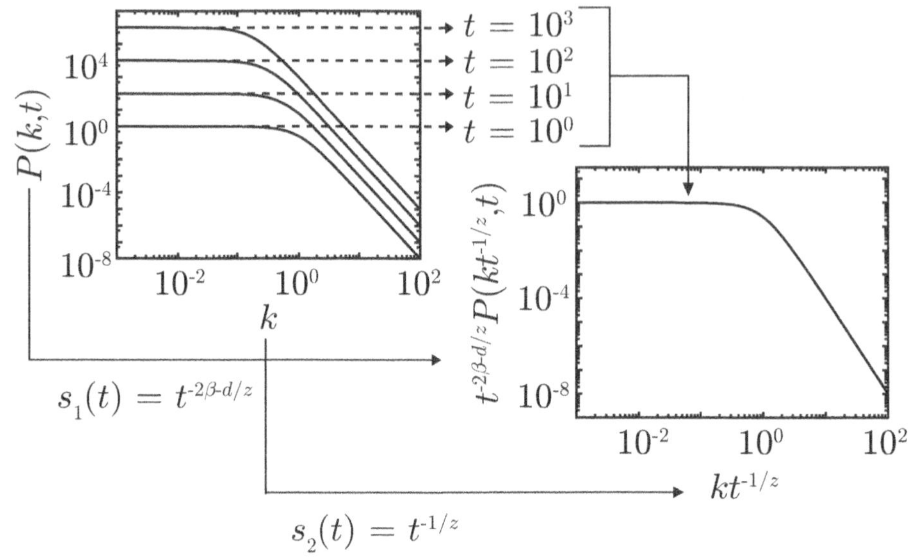

Fig. 3.5. Time-dependent scaling of a self-affine power spectral density function as described by (3.49). Scaling the horizontal axis by a factor $t^{-1/z}$ and the vertical axis by a factor $t^{-2\beta-d/z}$ collapses all curves onto one time-independent curve as given in (3.50).

a local and global length scale which scale with different exponents. Such a behavior can be observed when the local slope m evolves with a power law in time, $m(t) \sim t^{\kappa}$. So-called superrough surfaces [2, 23, 80] with a traditional roughness exponent $\alpha > 1$ have been described successfully by this theory. However, certain local models with $0 < \alpha < 1$ have also been described using anomalous scaling, including models utilizing random diffusion that describe fluid flow through porous materials [88, 89], and the Lai–Das Sarma–Villain equation, which describes growth in molecular beam epitaxy [68, 74].

3.6 Universality

From the discussion of the scaling properties of self-affine surfaces, the overall behavior of the surface can be summarized by the three scaling exponents α, β, and z. The specific details of the growth, such as the nature of the substrate, the source material, the deposition pressure and temperature, and numerous other factors did not contribute to the values of the growth exponents. This concept is known as universality.

The concept of universality is closely connected to scaling. It originated from the equilibrium statistical mechanical description of the collective behavior of a system near a critical point, a well-known example of which is the two-dimensional Ising system [150]. At the critical point, spin domains generated in the wild fluctuations of the system are present at all length scales, from very small scales to infinite size scales. The correlation function (called the spin–spin correlation in this context) scales and has the form $C(r) \sim r^{-\gamma}$, where $\gamma = \frac{1}{4}$. The system is self-similar, and the value of γ does not depend on the specific interaction energy between the spins. In fact, one observes similar behavior in other equivalent two-dimensional systems that may have nothing to do with spin. One example is a two-dimensional lattice gas system where occupied sites and empty sites correspond to spin up and spin down, respectively, as discussed in Wang and Lu [167]. Therefore, the value of the exponent γ is "universal".

The ideas of scaling and universality were then used to describe time-dependent dynamical systems, for example, the dynamics of an order–disorder phase transition of an alloy which is brought from a high-temperature disordered state quickly to a low-temperature ordered state where the order parameter is not conserved. The correlation function for this system can be written in a scaling form,

$$C(r,t) \sim g\left[\xi(t)\right] f\left(\frac{r}{\xi}\right), \tag{3.52}$$

where $\xi(t) \sim t^{\gamma}$ with $\gamma = \frac{1}{2}$ [147]. Again, the exponent does not depend on the material set and the microscopic interactions involved. This dynamic scaling concept was then used to formulate dynamic scaling theory in surface growth as presented in Sect. 3.5.

The growth exponents predicted by various continuum growth equations are said to comprise universality classes. The values for scaling exponents in 2+1 dimensions are given in Table 3.1. The exponents are obtained from a continuum equation of the form

$$\frac{\partial h(\mathbf{x},t)}{\partial t} = \Phi(\mathbf{x},t) + \eta(\mathbf{x},t), \tag{3.53}$$

where $\eta(\mathbf{x},t)$ represents the random noise that exists during growth. A more detailed discussion of some of these growth equations is given in Chap. 5. The study of nonlocal growth effects has led to a wide range of growth exponents that do not fall into specific universality classes, leading to a possible crossover effect from small β values ($\beta \leq 0.25$) to large β values ($\beta \approx 1$). In fact, for certain shadowing and reemission conditions, dynamic scaling breaks down and scaling relationships between the exponents cease to exist. However, the power-law scaling behavior associated with universality may still exist, which is discussed further in Chap. 8.

Table 3.1. Values for scaling exponents in various local growth models described by the continuum equation $\partial h/\partial t = \Phi + \eta$ in 2+1 dimensions ($d = 2$), where η is random noise. In the bulk diffusion model, j_z is the flux of atoms along the z direction, which is related to the chemical potential μ as $j_z \propto -\partial_z \mu$.

Φ	Equation	α	β	z	Reference
$\nu\nabla^2 h$	Edwards–Wilkinson	~ 0	0	2	[38]
$\nu\nabla^2 h + \frac{\lambda}{2}\vert\nabla h\vert^2$	KPZ	0.38	0.24	1.58	[12, 61]
$-\kappa\nabla^4 h$	Surface diffusion	1	$\frac{1}{4}$	4	[2, 25, 172]
$\Omega_0 j_z$	Bulk diffusion	0.5	0.2	3.33	[186]
$\nu\nabla^2 h - \kappa\nabla^4 h$		$0 - 1$	$0 - 0.25$	$2 - 4$	[99]
$-\kappa\nabla^4 h + \frac{\lambda}{2}\nabla^2\vert\nabla h\vert^2$	Lai–Das Sarma	$\frac{2}{3}$	$\frac{1}{5}$	$\frac{10}{3}$	[74]
$-\nu\nabla^2 h - \kappa\nabla^4 h + \frac{\lambda}{2}\vert\nabla h\vert^2$	KS (early time)	$0.75 - 0.80$	$0.22 - 0.25$	$3.0 - 4.0$	[33]
$-\nu\nabla^2 h - \kappa\nabla^4 h + \frac{\lambda}{2}\vert\nabla h\vert^2$	KS (late time)	$0.25 - 0.28$	$0.16 - 0.21$	–	[33]

4

Mounded Surfaces

In recent years, research interest has turned to understanding the dynamics of more complicated growth mechanisms that are characteristically nonlocal in nature. These investigations have been motivated by experimental results under certain types of deposition techniques including sputter deposition and chemical vapor deposition, most notably the measurement of growth exponents α, β, and z that are not consistent with the predictions of local growth models [8, 22, 35, 103, 143, 160, 185]. This is most evidently seen through an analysis of various values of the growth exponent β that have been reported in the literature for these deposition techniques, as shown in Fig. 4.1. In this figure, the spread of the majority of experimentally reported results is represented with a rectangle for each deposition technique, including thermal evaporation, sputter deposition, chemical vapor deposition, and oblique angle deposition. Most local models predict a relatively small value for β, as represented by the small spread of β for local models, which is evident from Table 3.1. Clearly, local models are not able to explain many of the experimental measurements of β. To explain these results, the theory of surface growth must be amended to include effects that can lead to such a wide range of experimental measurements, which invites the introduction of mounded surfaces.

When dealing with self-affine surfaces, there is only one lateral length scale, the lateral correlation length, beyond which surface heights are uncorrelated on the average. However, because self-affine surfaces have a unique scaling behavior, the magnitude of the lateral correlation length can be scaled to any arbitrary value, which implies that the lateral correlation length is not a true characteristic length scale of the surface, but rather a relative length scale. For example, in a self-affine surface morphology, there is no way to tell how "zoomed into" the surface you are looking, which is why scaling arguments hold for self-affine surfaces because zooming in with the right proportions yields a surface that is statistically identical to the original surface. This implies that there is no characteristic length scale on a self-affine surface because if there were, it would change upon zooming in, and the surface would

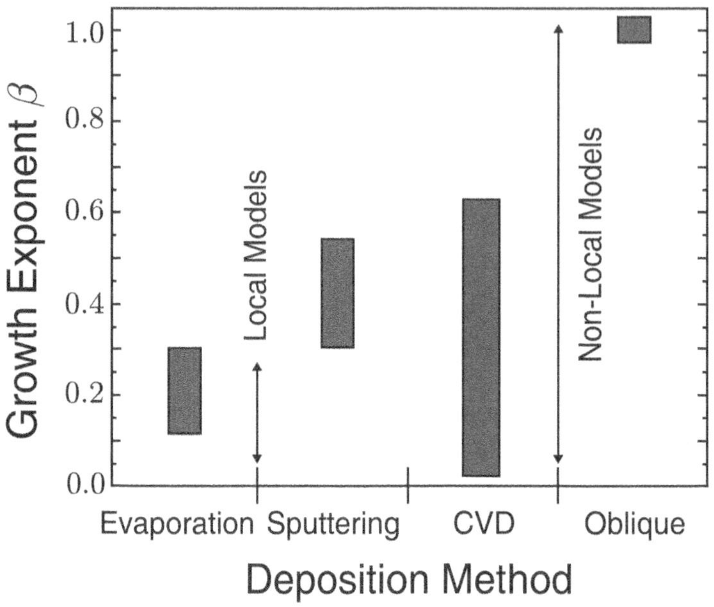

Fig. 4.1. In this plot of the growth exponent β, double-headed arrows indicate the range of β values predicted from both local and nonlocal models. Shaded areas represent a range of the majority of experimentally measured values for β reported in the literature for different deposition techniques.

no longer scale. It is possible for surfaces to possess a characteristic length scale, and such surfaces are called mounded surfaces. Clearly, from the above heuristic argument, mounded surfaces are not self-affine. This can be mathematically shown using the power spectral density function (PSD). If a surface possesses a characteristic length scale, it would result in a frequency peak in the PSD spectrum because the frequency corresponding to the characteristic length scale would be the most dominant in the surface profile. As a result, mounded surfaces are commonly defined as surfaces that have a characteristic peak in their PSD spectrum.

There are plenty of examples indicating the existence of mounded surfaces in experimental depositions, as well as in the etching of surfaces. Figure 4.2 shows PSD spectra from the surface topologies of (a) a Si film deposited by sputter deposition [122], (b) a Si film deposited by plasma chemical vapor deposition [22], and (c) a Si surface formed under plasma etching [32]. The inset of each graph is the AFM image from which the spectrum was measured. All surfaces exhibit a characteristic peak in the PSD, suggesting the existence of a characteristic length scale on the surface. As a result, none of these surfaces is self-affine. The validity of dynamic scaling is investigated for these

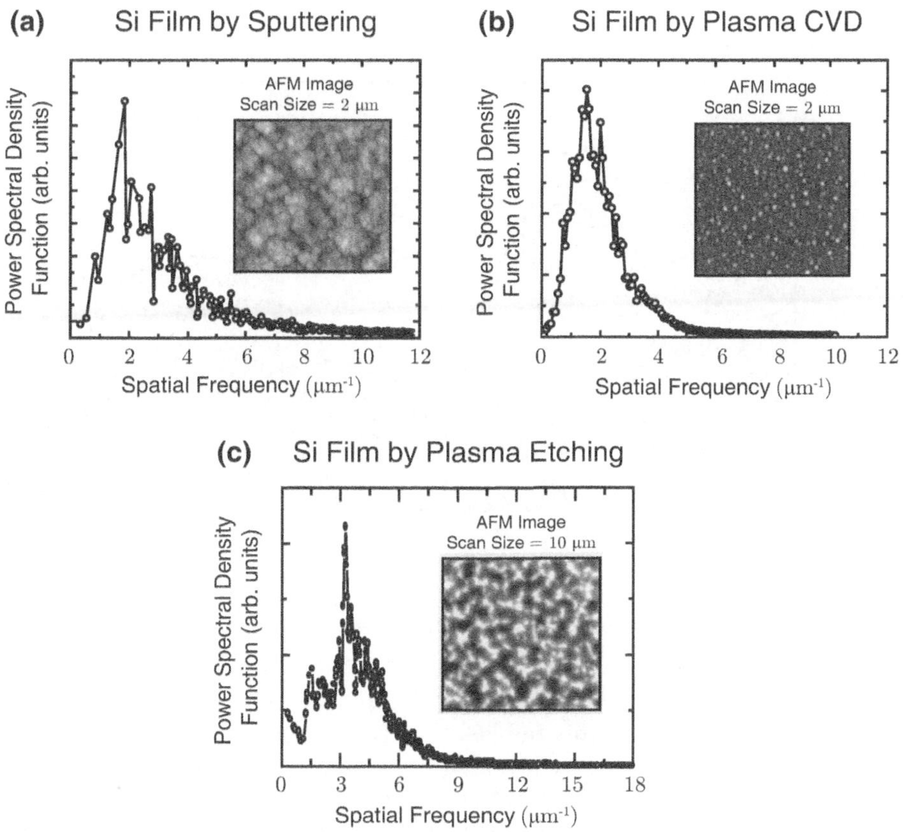

Fig. 4.2. Power spectral density (PSD) spectra of (**a**) a sputter deposited Si film [122]: thickness ≈ 6420 nm, RMS roughness ≈ 4 nm; (**b**) a plasma CVD Si film [22]: thickness ≈ 2250 nm, RMS roughness ≈ 6 nm; and (**c**) a plasma etched Si surface: etched thickness ≈ 6000 nm, RMS roughness ≈ 50 nm [185]. The insets are the corresponding AFM images of the surfaces. All PSD curves exhibit a characteristic peak, which implies that these surfaces are mounded.

surfaces and, in the case of sputtering and chemical vapor deposition, is shown to break down due to the mound formation. As described later, shadowing plays an important role in the generation of these mounds.

4.1 Length Scales λ and ξ

Even though there is a characteristic length scale for a mounded surface, called the wavelength λ, the lateral correlation length ξ is still well defined in terms of the autocorrelation function. The lateral correlation length was defined as the length beyond which surface heights were not significantly correlated.

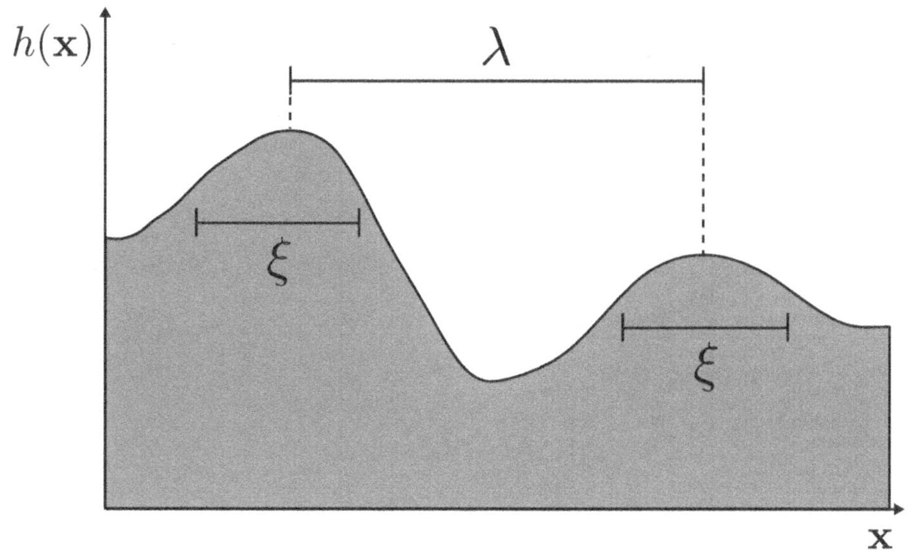

Fig. 4.3. Definition of the wavelength λ and the lateral correlation length ξ for a mounded surface. In general, the wavelength is not equal to the lateral correlation length, as seen in the figure.

For a mounded surface, this implies that the lateral correlation length is a measure of the size of the mounds. In some contexts, the lateral correlation length for a mounded surface is called the mound size, and denoted by ζ. The wavelength λ is related to the frequency peak in the PSD spectrum, and because the frequency peak in the PSD spectrum is a measure of the periodicity of mounds, this implies that the wavelength λ is a measure of the average distance between mounds. Note that the lateral correlation length ξ and the wavelength λ are defined differently and are not necessarily equal. They only must satisfy the relation $\xi \leq \lambda$ because mounds are separated by at least their size; only if mounds grow next to each other would it imply that $\xi = \lambda$. Figure 4.3 shows the definition of the lateral correlation length ξ and the wavelength λ for a 1+1 dimensional mounded surface.

4.2 Lateral Correlation Functions

The height–height correlation function for a mounded surface is similar in form to the height–height correlation function for a self-affine surface. The only notable difference in behavior arises at length scales beyond the lateral correlation length, or for $r > \xi$. In the self-affine case, the height–height correlation function is constant in this region, but for mounded surfaces it is oscillatory. This is a direct result of the characteristic peak in the PSD spectrum for mounded surfaces. A frequency peak implies that the surface profile

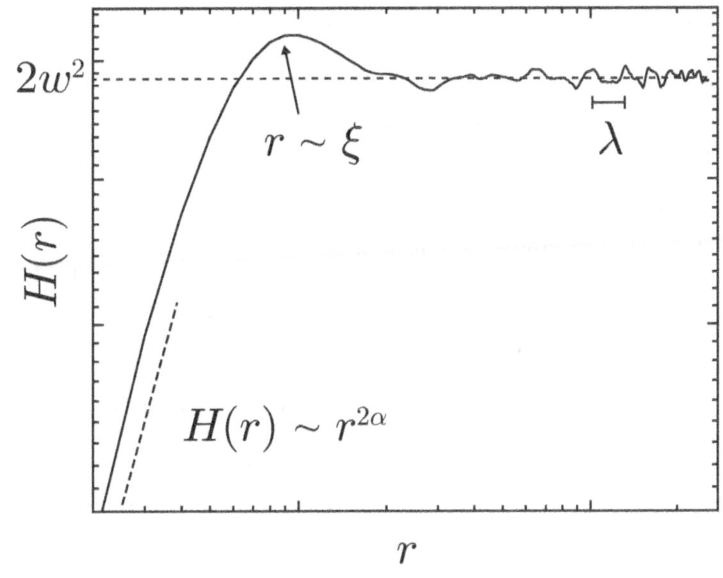

Fig. 4.4. Representative height–height correlation function for a mounded surface obtained from a simulated surface profile.

has a quasi-periodic behavior at the peak frequency, and this quasi-periodic behavior leads to oscillations in the height–height correlation function at large distances. One functional form for the height–height correlation function that behaves in this manner is given by, for a 2+1 dimensional surface [187],

$$H(r) = 2w^2 \left[1 - \exp\left[- \left(\frac{r}{\xi} \right)^{2\alpha} \right] J_0 \left(\frac{2\pi r}{\lambda} \right) \right], \tag{4.1}$$

where λ is the wavelength. This height–height correlation function is simply the exponential model for the height–height correlation function of a self-affine surface with an added oscillatory term to reflect wavelength selection. From this height–height correlation function, it follows that the autocorrelation function for a 2+1 dimensional mounded surface can be modeled as

$$R(r) = \exp\left[- \left(\frac{r}{\xi} \right)^{2\alpha} \right] J_0 \left(\frac{2\pi r}{\lambda} \right). \tag{4.2}$$

As was the case for self-affine surfaces, one can also introduce an autocorrelation function for mounded surfaces based on the K-correlation model [118], given by

$$R(r) = \frac{\alpha}{2^{\alpha-1}\Gamma(\alpha+1)} \left(\frac{r}{\xi}\sqrt{2\alpha} \right)^{\alpha} K_\alpha \left(\frac{r}{\xi}\sqrt{2\alpha} \right) J_0 \left(\frac{2\pi r}{\lambda} \right). \tag{4.3}$$

This autocorrelation function leads to an expression for the PSD that does not diverge as $\alpha \to 0$, whereas the exponential model breaks down in this limit. Also, the K-correlation model is able to give a rational expression for the PSD if $\alpha = 1$, whereas the exponential model gives a transcendental function that is more difficult to analyze. It should be noted that in 1+1 dimensions, the form of these autocorrelation functions is the same as in 2+1 dimensions except for the substitution of a cos function for the Bessel function J_0, which is required to obtain an analytic expression for the PSD in 1+1 dimensions as discussed in Sect. A.4.5.

In borrowing the self-affine height–height correlation function, the roughness exponent α carries over into the mounded height–height correlation function, which may seem contradictory. The roughness exponent α was defined in terms of the scaling behavior of a self-affine surface, and mounded surfaces are not self-affine, thus it would seem as if α would not have any meaning for mounded surfaces. However, recall that α reflects the short-range, or local roughness of a surface. On length scales much smaller than the wavelength λ, a mounded surface "appears" self-affine becuase there is no characteristic length scale smaller than the wavelength, and the local roughness is well defined in terms of the locally self-affine behavior of the mounded surface. In fact, from the form of the height–height correlation function, if $r \ll \lambda$, $J_0\left(2\pi r/\lambda\right) \approx 1$, and the self-affine height–height correlation function is recovered. Only when describing the long-range behavior ($r \geq \lambda$) of a mounded surface does the oscillatory term in the height–height correlation function become significant. Nevertheless, the local surface is not truly self-affine because scaling this surface beyond the wavelength destroys the self-affine scaling nature of the morphology.

An expression for the local slope of a mounded surface can be extracted from these model correlation functions using (3.21). The exponential model gives, for $\alpha = 1$,

$$m = \frac{w\sqrt{2}}{\xi}\sqrt{1 + \left(\frac{\pi\xi}{\lambda}\right)^2}. \tag{4.4}$$

However, for $\alpha < 1$, the mounded exponential model gives the same local slope as the self-affine exponential model. This occurs because the local slope depends only on the small r behavior of the autocorrelation function. The small r behavior of the exponential model for the autocorrelation function is

$$R(r) \approx 1 - \frac{r^{2\alpha}}{\xi^{2\alpha}} - \frac{\pi^2 r^2}{\lambda^2} + O\left(r^{2+2\alpha}\right). \tag{4.5}$$

When $\alpha < 1$, only the first two terms in this expansion are significant when evaluating the local slope because the term in r^2 is of too high an order. However, when $\alpha = 1$, both terms in r are of the same order, and both contribute to the value of the local slope. A similar result is obtained with the K-correlation model for the autocorrelation function, as a Taylor expansion will show similar small r behavior to (4.5). Therefore, according to this model, mounded behavior is only significant to the local slope when $\alpha = 1$; if $\alpha < 1$, the behavior of the surface at length scales smaller than the wavelength dominates in the measurement of the local slope. One should note that this result is a consequence of using these particular models for the autocorrelation function, and not a result that is necessarily true in general. For example, if the small r behavior of the autocorrelation function were modeled as

$$R(r) \approx 1 - \frac{r^{2\alpha}}{\xi^{2\alpha}} - \left(\frac{\pi r}{\lambda}\right)^{2\alpha} + O\left(r^{2+2\alpha}\right), \tag{4.6}$$

the local slope would become

$$m = \frac{(w\sqrt{2})^{1/\alpha}}{\xi}\left[1 + \left(\frac{\pi\xi}{\lambda}\right)^{2\alpha}\right]^{1/(2\alpha)} . \tag{4.7}$$

4.3 Power Spectral Density Function

Using the exponential model for the autocorrelation function of a mounded surface, the PSD in 2+1 dimensions can be modeled by, for $\alpha = 1$,

$$P(k) = \frac{w^2\xi^2}{4\pi}\exp\left[-\frac{(4\pi^2 + k^2\lambda^2)\xi^2}{4\lambda^2}\right] I_0\left(\frac{\pi k \xi^2}{\lambda}\right), \tag{4.8}$$

where $I_0(x)$ is the zeroth-order modified Bessel function of the first kind. It is noted that in some references [122, 187] this PSD differs by a factor of $(2\pi)^{-1}$, which is simply a matter of convention. The roughness exponent α is set equal to one in order to obtain a closed-form expression for the PSD, but when $\alpha \neq 1$, the PSD has the same characteristic shape as discussed in Sect. A.4. As introduced in (2.24), the peak position of the PSD is often found to behave as a power law in time, $k_m \sim t^{-p}$, where p is the wavelength exponent. This implies a similar behavior for the wavelength,

$$\lambda \sim t^p. \tag{4.9}$$

In addition, the full width at half maximum (FWHM) of the PSD of a mounded surface is inversely proportional to the lateral correlation length; FWHM $\propto \xi^{-1}$. A plot of the characteristic behavior of the PSD for a mounded surface is seen in Fig. 4.5.

The time-dependent scaling behavior of the PSD is related to the overall dynamic scaling behavior of the surface profile, as was shown in Sect. 3.5. In attempting to find scale factors $s_{1,2}(t)$ from (2.31) to remove the time-dependence from (4.8), we find that, in general, the time-dependence cannot be removed from the PSD. Recall that the parameters $w(t)$, $\xi(t)$, and $\lambda(t)$ all change with time and are not necessarily related. The scale factor $s_2(t)$ will not remove all the time-dependence from both the argument of the exponential and the Bessel function in general, regardless of the choice for $s_2(t)$. For

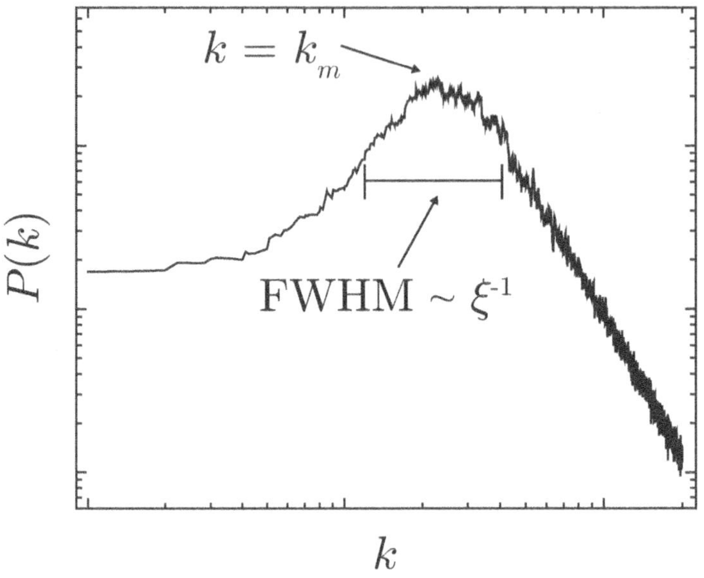

Fig. 4.5. Representative power spectral density function (PSD) for a mounded surface obtained from a simulated surface profile. The peak is located at $k = k_m$, and the full width at half maximum (FWHM) is inversely proportional to the lateral correlation length ξ.

example, choosing $s_2(t) = \xi^{-2}\lambda$ to remove the time-dependence from I_0, the argument of the exponential becomes

$$-\left(4\pi^2 + k^2\xi^{-4}\lambda^4\right)\frac{\xi^2}{4\lambda^2} = -\left(\frac{\pi^2\xi^2}{\lambda^2} + \frac{k^2\lambda^2}{4\xi^2}\right).$$

Because $\xi(t)$ and $\lambda(t)$ have a different time-dependence in general, the argument of the exponential still depends on time. However, in the special case where $\xi(t)$ and $\lambda(t)$ have the same time-dependence (i.e., $\xi(t) \propto \lambda(t)$), the ratio $\xi\lambda^{-1}$ is time-independent, and the PSD would simplify to, choosing $s_1(t) = 4\pi w^{-2}\xi^{-2}$,

$$Q(k) = \exp\left[-\left(\pi^2 + \frac{k^2}{4}\right)\right] I_0(\pi k),$$

which is independent of time. If we instead utilize the PSD from K-correlation model, as described in Sect. A.4.4, we would also find that the PSD exhibits time-dependent scaling only if $\xi(t) \propto \lambda(t)$. Thus, the PSD of a mounded surface only exhibits time-dependent scaling when $\xi(t) \propto \lambda(t)$, or, using the definitions of the time-dependent behaviors of the lateral correlation length and wavelength from (3.40) and (4.9), when $p = 1/z$. Because the time-dependent scaling of the PSD is a consequence of dynamic scaling, mounded surfaces should not obey dynamic scaling when $p \neq 1/z$. This point is investigated further in Chap. 8.

4.4 Origins of Mound Formation

The formation of mounds on a surface can be attributed to many different growth effects, and the most significant growth effects are discussed in the following sections. Growth effects that lead to mounds may be local or nonlocal in nature.

4.4.1 Step Barrier Diffusion Effect

There has been extensive research and examples of experiments [47, 55, 141, 152, 161, 162, 165, 166, 191, 190] performed on mounds formed by the step barrier diffusion effect during molecular beam epitaxy (MBE), also known as the Ehrlich–Schwoebel barrier effect. This effect does not allow atoms to diffuse over the edge of a step on the surface, which creates an overall uphill current of diffusive particle flux. This effect is a characteristically local growth effect because it involves the diffusion of particles on the surface. Diffusion, by definition, only affects atoms near the diffusing particle, and as a consequence the effects of diffusion are localized. However, the step barrier diffusion effect creates mounds on the surface, with the average mound separation λ evolving as a power law, $\lambda \sim t^p$, where the experimental values of the wavelength exponent p lie in the range from 0.16 to 0.26. In addition, the growth process can be modeled by a local stochastic continuum equation [55],

$$\frac{\partial h}{\partial t} = -\nu \nabla \cdot \left(\frac{\nabla h}{1 + |\nabla h|^2} \right) - \kappa \nabla^4 h + \eta, \qquad (4.10)$$

where the first term models the uphill growth due to the step barrier diffusion effect, and the second term is the Mullins diffusion term discussed in Sect. 5.1.4 to model the overall effect of surface diffusion. Note that this continuum equation involves only the height profile h and its derivatives, characteristic of the local nature of diffusion and the step barrier diffusion effect. Extensive studies have been carried out on the dynamic scaling behavior of surfaces grown under MBE. In general, dynamic scaling does not hold for these surfaces [113, 140], however under certain growth conditions it has been shown that dynamic scaling may hold [21, 24, 63, 117].

4.4.2 Shadowing

In many common thin film growth techniques such as sputter deposition and chemical vapor deposition (CVD), the growth dynamics are dominated by

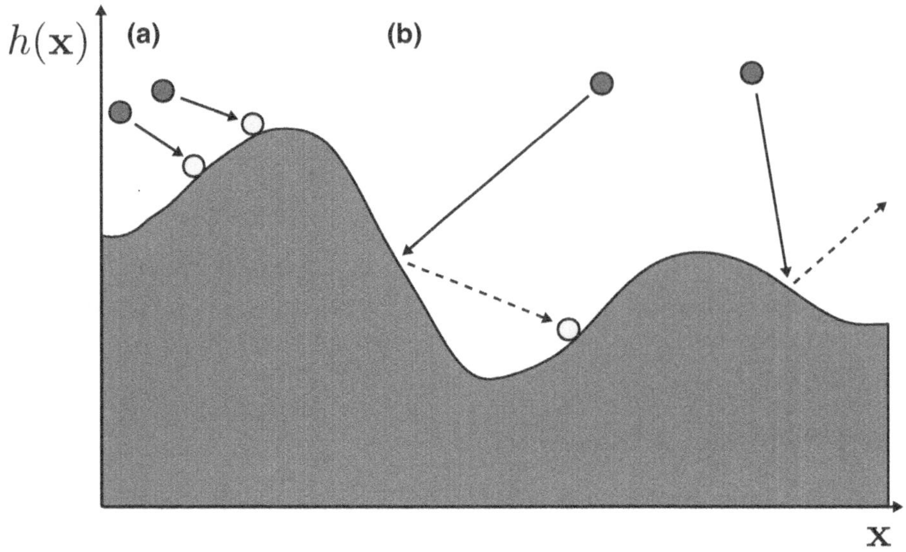

Fig. 4.6. Diagram of the nonlocal (**a**) shadowing effect and (**b**) reemission effect.

nonlocal growth effects. The primary nonlocal effect is the shadowing effect [62, 92, 177, 184], where taller surface features block incoming flux from reaching lower-lying areas of the surface. A diagram of the shadowing effect is seen in Fig. 4.6a. The shadowing effect is active because, in sputter deposition and CVD, the incoming flux has an angular distribution. This allows taller surface features to grow at the expense of shorter ones, leading to a competition between different surface features for particle flux. This competition ultimately leads to a mounded surface as shorter surface features receive little or no particle flux and "die out". Shadowing is an inherently nonlocal process because the shadowing of a surface feature depends on the heights of all other surface features, not just close, or local, ones.

4.4.3 Reemission

In addition, the formation of mounds due to the shadowing effect can be hindered by the reemission of particles during deposition. The reemission effect allows particles to "bounce around" before they settle at appropriate sites on the surface [184]. A diagram of the reemission effect is seen in Fig. 4.6b. Reemitted particles serve to change the overall particle flux incident on the surface, allowing previously shadowed surface features to receive particle flux. To describe the reemission effect, a sticking coefficient (s_0) is used which represents the probability that a particle will stick to the surface when it first strikes. Higher-order sticking coefficients ($s_{n>0}$) represent the probability that a particle will stick having been reemitted n times. During deposition,

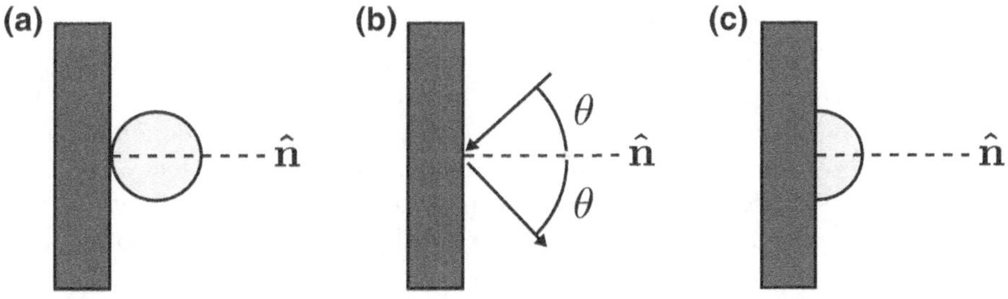

Fig. 4.7. Polar probability plots of reemitted flux distributions for (**a**) thermal reemission, (**b**) specular reemission, and (**c**) uniform reemission [32].

shadowing tends to roughen the surface and reemission tends to smooth the surface [92]. Thus, growth under perfect shadowing would correspond to no reemission ($s_0 = 1$).

When considering the reemission effect, one must not only specify the value of the sticking coefficients, but also the mode of reemission [32, 35]. For example, when a particle is reemitted, the direction it assumes after reflection may or may not depend on its incident direction. Different types of reemission are depicted in Fig. 4.7. Thermal reemission assumes that, once the particle comes in contact with the surface, it attains a thermal equilibrium with the surface, and reflects off with a Maxwellian distribution of velocities. In particular, a particle with velocity $\mathbf{v}'$ has a probability $\mathcal{P}(\mathbf{v}', \mathbf{v})$ of attaining a velocity $\mathbf{v}$ after thermal reemission given by

$$\mathcal{P}(\mathbf{v}', \mathbf{v}) = \frac{\mathbf{v} \cdot \hat{\mathbf{n}}}{2\pi\theta^2} \exp\left(-\frac{v^2}{2\theta}\right), \tag{4.11}$$

where $\theta = kT/m$, k is the Boltzmann constant, T is the surface temperature, m is the mass of the particle, and $\hat{\mathbf{n}}$ is the surface normal at the impact point. This model for reemission is depicted in Fig. 4.7a. Notice that the reemitted velocity does not depend on the incident velocity, the reemission is diffuse. This model may be applicable when the local surface roughness is significantly larger than the size of the particle. In this case, the reemitted velocity will depend highly on the surface profile at the impact point, which varies considerably over the surface, and an ensemble average will give the behavior reflected in (4.11).

Another model for reemission is called specular reemission, where the particle reflects off the surface as if it were a billiard ball striking a smooth surface. For specular reemission, the probability that a particle with velocity $\mathbf{v}'$ is reemitted with velocity $\mathbf{v}$ is given by

$$R(\mathbf{v}', \mathbf{v}) = \delta(\mathbf{v} - \mathbf{v}' + 2\hat{\mathbf{n}}(\hat{\mathbf{n}} \cdot \mathbf{v}')). \tag{4.12}$$

The case of specular reemission is pictured in Fig. 4.7b. As opposed to thermal reemission, specular reemission would be valid when the particle size is significantly larger than the surface roughness, and the reflected velocity depends only on the direction of the incident velocity and the local surface normal. A third model for reemission, uniform reemission, assumes that particles are reflected with velocities uniformly distributed in θ, and is pictured in Fig. 4.7c. The specific type of reemission used in a growth model depends on the specific deposition conditions to be modeled, as well as efficiency considerations because it is more straightforward to compute the result of specular reemission for a particle in a discrete model than thermal reemission.

Part II

Continuum Surface Growth Models

5

Stochastic Growth Equations

The first class of growth models we consider are growth models based on continuum growth equations, also known as Langevin equations. These models are often able to predict values for the exponents α, β, and z analytically, and form the basis for the universality classes introduced in Sect. 3.6. The general form of a stochastic continuum equation is [8, 64]

$$\frac{\partial h(\mathbf{x},t)}{\partial t} = \Phi\left(\mathbf{x},\{h\},t\right) + \eta(\mathbf{x},t), \tag{5.1}$$

where $\eta(\mathbf{x},t)$ is the noise in the system, often assumed to be Gaussian, which satisfies the properties

$$\langle \eta(\mathbf{x},t)\rangle = 0 \quad \text{and} \quad \langle \eta(\mathbf{x},t)\eta(\mathbf{x}',t')\rangle = 2D\delta^d(\mathbf{x}-\mathbf{x}')\delta(t-t'), \tag{5.2}$$

and $\Phi\left(\mathbf{x},\{h\},t\right)$ is some function of the height profile that reflects the growth processes to be modeled. The function $\Phi\left(\mathbf{x},\{h\},t\right)$ can take on many forms, and the most commonly used forms are discussed in the following sections.

5.1 Local Models

We first consider local continuum models, where the function Φ depends on local interaction terms only, of which derivatives are the most common. Models that include nonlocal effects build off the results of local models.

5.1.1 Random Deposition

The simplest growth process that can be modeled using a stochastic continuum equation is the process where $\Phi\left(\mathbf{x},\{h\},t\right)$ equals a constant, C. This implies that there is no growth process active to correlate surface heights. The mean height of the surface evolves as

$$\frac{\partial \overline{h}}{\partial t} = \frac{\partial}{\partial t}\langle h(\mathbf{x},t)\rangle = \left\langle \frac{\partial h(\mathbf{x},t)}{\partial t} \right\rangle = \langle C + \eta(\mathbf{x},t)\rangle = C. \tag{5.3}$$

The average can be interchanged with the derivative because the average is an integral over position, not time. Thus, the mean height grows as $\overline{h} = Ct$. The interface width, which can be expressed as

$$[w(t)]^2 = \left\langle [h(\mathbf{x},t)]^2 \right\rangle - \overline{h}^2,$$

can also be explicitly computed. A formal expression for $h(\mathbf{x},t)$ is found by integrating the continuum equation in time,

$$h(\mathbf{x},t) = \int_0^t \left(\frac{\partial h(\mathbf{x},t')}{\partial t'} \right) \mathrm{d}t' = \int_0^t C\mathrm{d}t' + \int_0^t \eta(\mathbf{x},t')\mathrm{d}t' = Ct + \int_0^t \eta(\mathbf{x},t')\mathrm{d}t'. \tag{5.4}$$

It follows that

$$\begin{aligned}
\left\langle [h(\mathbf{x},t)]^2 \right\rangle &= \left\langle \left[Ct + \int_0^t \eta(\mathbf{x},t')\mathrm{d}t' \right]^2 \right\rangle \\
&= \left\langle (Ct)^2 + 2Ct\int_0^t \eta(\mathbf{x},t')\mathrm{d}t' \right. \\
&\quad \left. + \left(\int_0^t \eta(\mathbf{x},t')\mathrm{d}t' \right) \left(\int_0^t \eta(\mathbf{x},t'')\mathrm{d}t'' \right) \right\rangle \\
&= (Ct)^2 + 2Ct\int_0^t \langle \eta(\mathbf{x},t')\rangle \mathrm{d}t' + \int_0^t\int_0^t \langle \eta(\mathbf{x},t')\eta(\mathbf{x},t'')\rangle \mathrm{d}t'\mathrm{d}t'' \\
&= (Ct)^2 + 2Ct\int_0^t (0)\mathrm{d}t' + \int_0^t\int_0^t 2D\delta(t'-t'')\mathrm{d}t'\mathrm{d}t'' \\
&= (Ct)^2 + 2Dt,
\end{aligned}$$

and the interface can be expressed as

$$[w(t)]^2 = \left\langle [h(\mathbf{x},t)]^2 \right\rangle - \overline{h}^2 = (Ct)^2 + 2Dt - (Ct)^2 = 2Dt \sim t, \tag{5.5}$$

which gives $\beta = \frac{1}{2}$. However, because the heights are not correlated, the lateral correlation length ξ is always zero, and the dynamic exponent z is not defined. Also, because the interface width does not saturate due to the lack of correlation, α is also not defined and, as a result, the surface is not self-affine. As such, the random deposition model does not completely describe any realistic experiment, but does serve as an analytically solvable model with an exact prediction of $\beta = \frac{1}{2}$, which is often observed at very early times during growth from a flat substrate when noise is the most dominant growth mechanism.

5.1.2 Edwards–Wilkinson Equation

When surface heights are correlated, the random deposition model is no longer valid, and $\Phi(\mathbf{x}, \{h\}, t)$ must be modified to include correlations between surface heights. Before more complicated nonequilibrium growth models are considered, it is beneficial to first deduce symmetries that a surface may satisfy [8], and build off these ideas to formulate more complicated growth models. One such symmetry is the independence of the definition of the origin of the coordinate system, or the origin of time, which implies invariance under the transformations

$$h \to h + \Delta h \tag{5.6}$$
$$\mathbf{x} \to \mathbf{x} + \Delta \mathbf{x} \tag{5.7}$$
$$t \to t + \Delta t. \tag{5.8}$$

The surface should also be symmetric about the origin of the coordinate system, as well as the mean height, which is taken always to equal zero by a choice of reference height, which gives invariance under the transformations

$$\mathbf{x} \to -\mathbf{x} \tag{5.9}$$
$$h \to -h. \tag{5.10}$$

Taking these symmetry arguments into account, the lowest-order term that satisfies these symmetries is the Laplacian of h, $\nabla^2 h$. The growth equation involving this term is called the Edwards–Wilkinson (EW) equation, and is given by [38]

$$\frac{\partial h}{\partial t} = \nu \nabla^2 h + \eta, \tag{5.11}$$

where the Laplacian term in the EW equation is referred to as the surface relaxation term, because the effect of the Laplacian is to smooth the surface profile while keeping the mean height unchanged. The exponents α, β, and z can be obtained using a scaling argument, rescaling the variables $\mathbf{x} \to \varepsilon \mathbf{x}$, $h \to \varepsilon^\alpha h$, and $t \to \varepsilon^z t$, which gives

$$\frac{\partial (\varepsilon^\alpha h)}{\partial (\varepsilon^z t)} = \nu \nabla^2 (\varepsilon^\alpha h) + \eta (\varepsilon \mathbf{x}, \varepsilon^z t).$$

Using the definition of the noise $\eta(\mathbf{x}, t)$,

$$\begin{aligned} \langle \eta(\varepsilon \mathbf{x}, \varepsilon^z t) \eta(\varepsilon \mathbf{x}', \varepsilon^z t') \rangle &= 2D\delta^d(\varepsilon(\mathbf{x} - \mathbf{x}'))\delta(\varepsilon^z(t - t')) \\ &= 2D\varepsilon^{-(d+z)}\delta^d(\mathbf{x} - \mathbf{x}')\delta(t - t'), \end{aligned}$$

because $\delta^d(\varepsilon \mathbf{x}) = \varepsilon^{-d}\delta^d(\mathbf{x})$. This implies that

$$\eta(\varepsilon \mathbf{x}, \varepsilon^z t) \to \varepsilon^{-(d+z)/2}\eta(\mathbf{x}, t).$$

Thus, the scaled equation becomes

$$\begin{aligned}\varepsilon^{\alpha-z}\frac{\partial h}{\partial t} &= \varepsilon^{\alpha-2}\nu\nabla^2 h + \varepsilon^{-(d+z)/2}\eta(\mathbf{x},t)\\ \frac{\partial h}{\partial t} &= \varepsilon^{z-2}\nu\nabla^2 h + \varepsilon^{-(d-z)/2-\alpha}\eta(\mathbf{x},t).\end{aligned}$$

In order to preserve scale invariance, this equation must be identical to (5.11), which gives

$$z = 2;\ \ \alpha = \frac{2-d}{2};\ \ \beta = \frac{\alpha}{z} = \frac{2-d}{4}. \tag{5.12}$$

These exponents characterize the growth of a surface under the EW equation. There is clearly a problem when this argument is applied to surfaces with $d \geq 2$, and the $d = 2$ case can discussed in terms of the behavior of the power spectral density function.

With the form of the EW equation given in (5.11), we can find an analytic expression for the power spectral density function (PSD) of a surface that evolves under these growth dynamics. We can define the Fourier transform of the surface height $\hat{h}(\mathbf{k},t)$ as

$$\hat{h}(\mathbf{k},t) \equiv \frac{1}{(2\pi)^{d/2}}\int h(\mathbf{x},t)\mathrm{e}^{-\mathrm{i}\mathbf{k}\cdot\mathbf{x}}\mathrm{d}\mathbf{x}. \tag{5.13}$$

If we multiply the EW equation by $(2\pi)^{-d/2}\mathrm{e}^{-\mathrm{i}\mathbf{k}\cdot\mathbf{x}}$, integrate over $\mathbf{x}$, and use the chain rule to integrate over the Laplacian term, we obtain a differential equation for $\hat{h}(\mathbf{k},t)$,

$$\frac{\partial\hat{h}(\mathbf{k},t)}{\partial t} = -\nu k^2\hat{h}(\mathbf{k},t) + \Theta(\mathbf{k},t), \tag{5.14}$$

where $\Theta(\mathbf{k},t)$ is the Fourier transform of the noise,

$$\Theta(\mathbf{k},t) = \frac{1}{(2\pi)^{d/2}}\int \eta(\mathbf{x},t)\mathrm{e}^{-\mathrm{i}\mathbf{k}\cdot\mathbf{x}}\mathrm{d}\mathbf{x}. \tag{5.15}$$

If we take an ensemble average over time, the properties of $\Theta(\mathbf{k},t)$ are similar to the properties of $\eta(\mathbf{x},t)$ given in (5.2),

$$\langle\Theta(\mathbf{k},t)\rangle = \frac{1}{(2\pi)^{d/2}}\int\langle\eta(\mathbf{x},t)\rangle\mathrm{e}^{-\mathrm{i}\mathbf{k}\cdot\mathbf{x}}\mathrm{d}\mathbf{x} = 0, \tag{5.16}$$

$$\begin{aligned}\langle\Theta(\mathbf{k},t)\Theta(\mathbf{k}',t')\rangle &= \frac{1}{(2\pi)^d}\int\int\langle\eta(\mathbf{x},t)\eta(\mathbf{x}',t')\rangle\mathrm{e}^{-\mathrm{i}\mathbf{k}\cdot\mathbf{x}}\mathrm{e}^{-\mathrm{i}\mathbf{k}'\cdot\mathbf{x}'}\mathrm{d}\mathbf{x}\mathrm{d}\mathbf{x}'\\ &= \frac{1}{(2\pi)^d}\int\int\left(2D\delta^d(\mathbf{x}-\mathbf{x}')\delta(t-t')\right)\mathrm{e}^{-\mathrm{i}\mathbf{k}\cdot\mathbf{x}}\mathrm{e}^{-\mathrm{i}\mathbf{k}'\cdot\mathbf{x}'}\mathrm{d}\mathbf{x}\mathrm{d}\mathbf{x}'\\ &= \frac{2D\delta(t-t')}{(2\pi)^d}\int\mathrm{e}^{-\mathrm{i}(\mathbf{k}+\mathbf{k}')\cdot\mathbf{x}}\mathrm{d}\mathbf{x}\\ &= 2D\delta(\mathbf{k}+\mathbf{k}')\delta(t-t').\end{aligned} \tag{5.17}$$

The differential equation (5.14) is a first-order differential equation in time, and can be solved using an integrating factor to give the solution

$$\hat{h}(\mathbf{k},t) = \mathrm{e}^{-\nu k^2 t}\left[\int_0^t \Theta(\mathbf{k},t')\mathrm{e}^{\nu k^2 t'}\mathrm{d}t' + C\right]. \tag{5.18}$$

If we begin from a flat surface, $h(\mathbf{x},0) = 0$, then $\hat{h}(\mathbf{k},0) = 0$ which gives $C = 0$ and

$$\hat{h}(\mathbf{k},t) = \mathrm{e}^{-\nu k^2 t}\int_0^t \Theta(\mathbf{k},t')\mathrm{e}^{\nu k^2 t'}\mathrm{d}t'. \tag{5.19}$$

Using the definition of the power spectral density function given in (2.15), and denoting the complex conjugate of $\Theta(\mathbf{k},t)$ by $\Theta^*(\mathbf{k},t)$,

$$\begin{aligned}
P(\mathbf{k},t) &= \left|\hat{h}(\mathbf{k},t)\right|^2 \\
&= \mathrm{e}^{-2\nu k^2 t}\int_0^t\int_0^t \Theta(\mathbf{k},t')\Theta^*(\mathbf{k},t'')\mathrm{e}^{\nu k^2 t'}\mathrm{e}^{\nu k^2 t''}\mathrm{d}t'\mathrm{d}t'' \\
&= 2D\mathrm{e}^{-2\nu k^2 t}\int_0^t \mathrm{e}^{2\nu k^2 t'}\mathrm{d}t' \\
&= 2D\mathrm{e}^{-2\nu k^2 t}\left[\frac{\mathrm{e}^{2\nu k^2 t}-1}{2\nu k^2}\right] \\
&= D\frac{1-\mathrm{e}^{-2\nu k^2 t}}{\nu k^2}.
\end{aligned} \tag{5.20}$$

From the scaling argument given in (5.12), $d = 2$ is the critical dimension of the EW equation, as the scaling argument predicts $\alpha = \beta = 0$ for $d = 2$, which suggests that the behavior of the roughness is more complicated than a power law. To determine the behavior of the interface width in 2+1 dimensions, we can use the relation from (2.17),

$$w^2 = \int P(\mathbf{k},t)\mathrm{d}\mathbf{k} = \frac{2\pi D}{\nu}\int_0^\infty \frac{1-\mathrm{e}^{-2\nu k^2 t}}{k}\mathrm{d}k. \tag{5.21}$$

Unfortunately, this integral does not converge, which occurs because any real surface can only exhibit self-similar behavior up to a cutoff length scale a, and the EW equation does not represent the growth dynamics at length scales below this scale. If there is a lower bound on the length scales involved in the problem, then there is a similar upper bound on the frequency scales involved in the problem. This implies that the PSD derived from the EW equation is only valid in the domain $k \in [0, a^{-1}]$, and the interface width behaves as

$$w^2 \propto \int_0^{a^{-1}} \frac{1-\mathrm{e}^{-2\nu k^2 t}}{k}\mathrm{d}k. \tag{5.22}$$

However, from the argument of the asymptotic behavior of a general PSD given in Sect. 3.4, a cutoff in the integral can be approximated by an appropriate exponential to aid in the evaluation of the integral,

$$w^2 \propto \int_0^\infty \frac{1 - \mathrm{e}^{-2\nu k^2 t}}{k} \mathrm{e}^{-a^2 k^2} \mathrm{d}k = \int_0^\infty \frac{\mathrm{e}^{-a^2 k^2} - \mathrm{e}^{-(2\nu t + a^2) k^2}}{k} \mathrm{d}k. \tag{5.23}$$

With a change of variable $t = k^2$, this relation becomes

$$w^2 \propto \int_0^\infty \frac{\mathrm{e}^{-a^2 t} - \mathrm{e}^{-(2\nu t + a^2) t}}{t} \mathrm{d}t = \ln \frac{2\nu t + a^2}{a^2} = \ln \left(1 + \frac{2\nu t}{a^2}\right), \tag{5.24}$$

where the integral was evaluated using [149]. Therefore, the roughness behaves as

$$w(t) \sim \sqrt{\ln \left(1 + \frac{2\nu t}{a^2}\right)}, \tag{5.25}$$

which is not in the form $w(t) \sim t^\beta$. Note that for early times, $t \ll a^2/(2\nu)$, because $\ln(1 + x) \approx x$ for small x, this expression gives $\beta = \frac{1}{2}$, which is consistent with the random deposition model. However, for long times, we observe a logarithmic behavior for the roughness. A logarithmic behavior could have been expected from the prediction that $\beta = 0$ for $d = 2$ from the simple scaling argument, which can be written as

$$w^2 \sim t^{2\beta} = \exp\left[2\beta \ln t\right] \approx 1 + 2\beta \ln t + O\left((2\beta \ln t)^2\right), \tag{5.26}$$

for the domain where $|\ln t| \ll (2\beta)^{-1}$, which would be significant if β were very small. Similarly, a prediction of $\alpha = 0$ implies a logarithmic behavior for the small r behavior of the height–height correlation function,

$$H(r) \sim \log r, \quad r \ll \xi. \tag{5.27}$$

5.1.3 Kardar–Parisi–Zhang Equation

The symmetries exhibited by the EW equation may be broken, in particular the statement that height fluctuations are symmetric with respect to the mean height as growth can occur along the local surface normal, which clearly violates this symmetry. If growth along the local surface normal occurs at a rate v, then in a time Δt the change in vertical height Δh of the surface is given by

$$\Delta h = \sqrt{(v\Delta t)^2 + (v\Delta t\, |\nabla h|)^2} = v\Delta t \sqrt{1 + |\nabla h|^2} = v\Delta t \left(1 + \frac{|\nabla h|^2}{2} + \cdots\right), \tag{5.28}$$

when the local slope is small, $|\nabla h| \ll 1$. According to this derivation, the EW growth equation can be amended to include growth along the local surface normal,

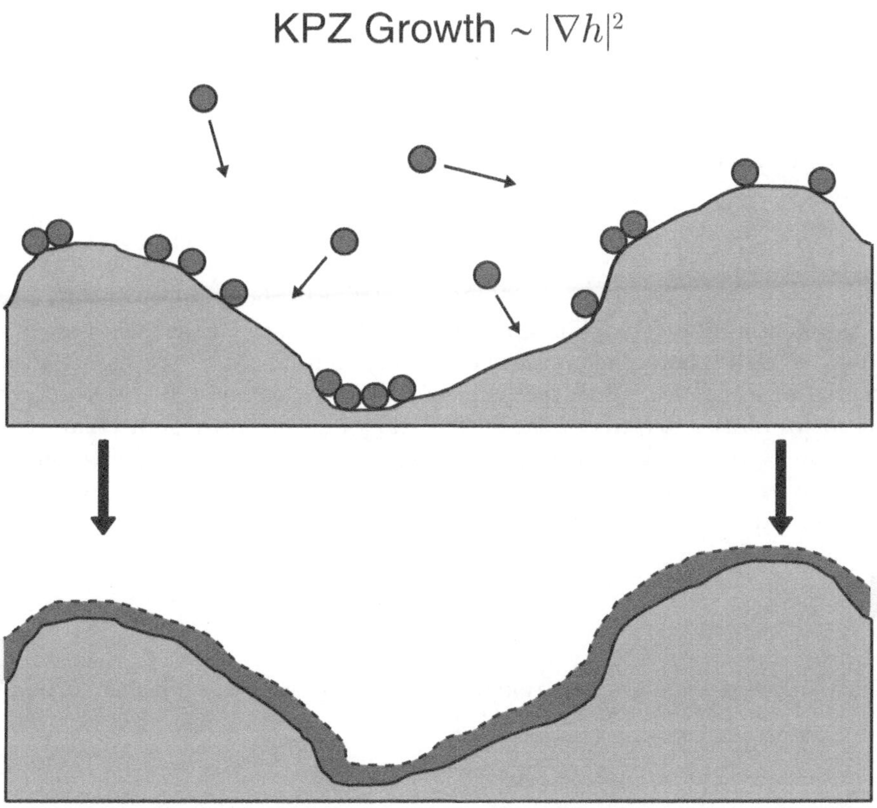

Fig. 5.1. The effect of the KPZ term $|\nabla h|^2$ on a surface profile. Because the growth occurs along the surface normal, the growth is conformal.

$$\frac{\partial h}{\partial t} = \nu \nabla^2 h + \frac{\lambda}{2} |\nabla h|^2 + \eta. \tag{5.29}$$

This equation is known as the Kardar–Parisi–Zhang (KPZ) equation [61]. A diagram of the growth dynamics modeled by the KPZ equation is included in Fig. 5.1. To obtain the exponents for this growth equation, the simple scaling argument used with the EW equation is no longer valid because the constants ν and λ in the KPZ equation and the constant D in the correlation of the noise do not all rescale independently. Renormalization group theory can be used to obtain the exponents, which gives exact values only in 1+1 dimensions,

$$z = \frac{3}{2}; \quad \alpha = \frac{1}{2}; \quad \beta = \frac{\alpha}{z} = \frac{1}{3}. \tag{5.30}$$

The exponents in 2+1 dimensions have only been investigated using simulations, with the result [12]

$$z = 1.58; \quad \alpha = 0.38; \quad \beta = 0.24. \tag{5.31}$$

5.1.4 Mullins Diffusion Equation

To model surface diffusion in a stochastic continuum equation, consider a macroscopic current of particles on the surface, represented by the vector $\mathbf{j}(\mathbf{x}, t)$. Because diffusion conserves the total number of particles on the surface, $\mathbf{j}(\mathbf{x}, t)$ must satisfy the continuity relation [8],

$$\frac{\partial h(\mathbf{x}, t)}{\partial t} = -\nabla \cdot \mathbf{j}(\mathbf{x}, t).$$

In addition, the surface current $\mathbf{j}(\mathbf{x}, t)$ is related to the gradient of the chemical potential, $\mathbf{j}(\mathbf{x}, t) \propto -\nabla \mu(\mathbf{x}, t)$, because the surface current will flow from areas of higher potential to areas of lower potential. Also, the chemical potential $\mu(\mathbf{x}, t)$ is related to the number of bonds that must be broken by an atom to diffuse. Regions of the surface that have a positive curvature have more available bonds, which in turn makes it harder for an atom to diffuse. Conversely, regions of the surface with negative curvature have fewer available bonds, and an atom can diffuse more readily. These conditions are satisfied if $\mu(\mathbf{x}, t) \propto -\nabla^2 h(\mathbf{x}, t)$. Combining these results,

$$\frac{\partial h(\mathbf{x}, t)}{\partial t} = -\nabla \cdot \mathbf{j}(\mathbf{x}, t) = -\nabla \cdot \left[-\nabla(-\kappa \nabla^2 h(\mathbf{x}, t))\right] = -\kappa \nabla^4 h(\mathbf{x}, t).$$

This suggests adding a biharmonic term to the growth equation to model surface diffusion,

$$\frac{\partial h}{\partial t} = -\kappa \nabla^4 h + \eta. \tag{5.32}$$

This is known as the Mullins diffusion equation [2, 25, 114, 172]. The effect of the Mullins diffusion equation on a surface profile is pictured in Fig. 5.2. A scaling argument similar to the argument used to obtain the scaling exponents in the EW equation can be used to obtain

$$z = 4; \quad \alpha = \frac{4-d}{2}; \quad \beta = \frac{\alpha}{z} = \frac{4-d}{8}. \tag{5.33}$$

In addition, the PSD of a surface evolving under the Mullins diffusion equation can be found by a procedure similar to the derivation of the PSD of the EW equation in Sect. 5.1.2 to give

$$P(k, t) = D\frac{1 - \mathrm{e}^{-2\kappa k^4 t}}{\kappa k^4}. \tag{5.34}$$

Often, the Mullins diffusion term is added to the KPZ equation when surface diffusion is active. Experimental investigations into growth dominated by surface diffusion, as described by the Mullins diffusion equation, suggest that the growth is nonstationary [54, 91]; that is, the local slope m changes with time as

$$m(t) \sim \sqrt{\ln t}. \tag{5.35}$$

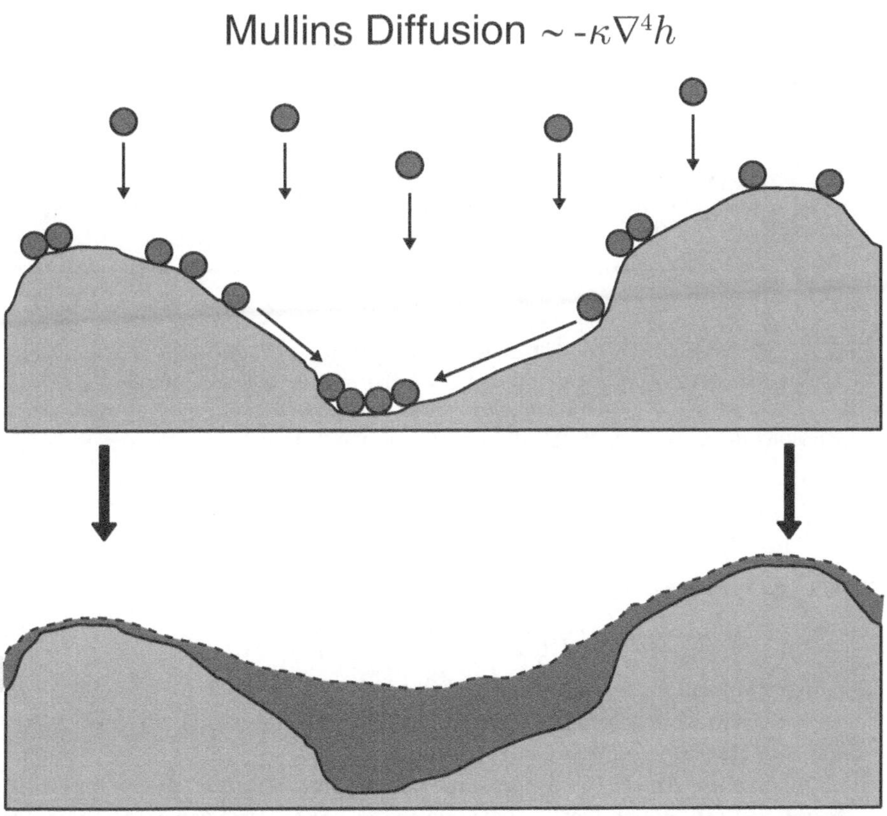

Fig. 5.2. The effect of the Mullins diffusion term $-\kappa\nabla^4 h$ on a surface profile. Recall that this term describes growth from a frame with zero mean height, which leads to growth in low-lying, large curvature areas of the surface.

The influence of a time-dependent local slope on the height–height correlation function is shown in Fig. 5.3. When the local slope is constant in time, height–height correlation functions coincide for $r \ll \xi$, but differ when the local slope m changes in time. Recall that the local slope m is related to the roughness and correlation length as

$$m \sim \frac{w^{1/\alpha}}{\xi}.$$

The time-dependence of the local slope can be expressed in terms of the roughness as, using the dynamic scaling relation $\beta = \alpha/z$,

$$w(t) \sim (m\xi)^{\alpha} \sim t^{\beta} \left[\ln t\right]^{\alpha/2}. \tag{5.36}$$

However, if we plot the interface width as a function of time on a log–log scale to measure β, we will obtain the curve

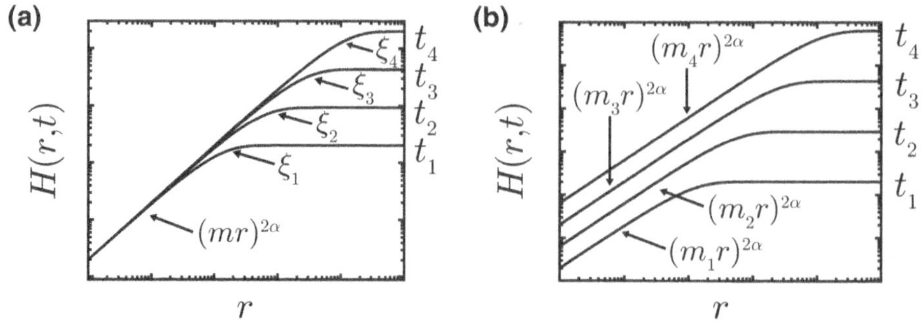

Fig. 5.3. Diagram of the height–height correlation function under (**a**) a stationary local slope m that is constant in time, and (**b**) a nonstationary local slope m that changes with time. A nonstationary local slope has been observed in experimental depositions described by the Mullins diffusion equation [91].

$$\ln w \sim \beta \ln t + \frac{\alpha}{2} \ln \ln t,$$

which has a slope

$$\frac{\mathrm{d}(\ln w)}{\mathrm{d}(\ln t)} \sim \beta \left[1 + \frac{\alpha}{2\beta \ln t}\right]. \tag{5.37}$$

Because the local slope shows a logarithmic behavior for long times, $t \gg 1$, in measuring experimental data it is difficult to pick up the $(\ln t)^{-1}$ term when measuring the slope, and the data would suggest a value for β consistent with dynamic scaling. In $d = 2$ dimensions, (5.33) predicts that $\beta = \frac{1}{4}$ for Mullins diffusion growth, which would be the value for β measured from a log–log plot of the interface width against time. This behavior is reasonable considering the behavior of the roughness under the EW equation, which showed a similar logarithmic dependence in two dimensions. Since $\beta = 0$ in two dimensions under the EW equation, the logarithm is all that is significant in (5.36), and the logarithmic dependence can be explicitly seen. Mullins diffusion has $\beta \neq 0$ in two dimensions, therefore the logarithm gets "hidden" by the power law in (5.36), as was argued in Sect. 3.5.

5.2 Nonlocal Models

A continuous model for shadowing was introduced by Karunarisiri et al., which included a term proportional to the "solid exposure angle" Ω at each point on the surface [34, 62, 177, 178],

$$\frac{\partial h}{\partial t} = -\kappa \nabla^4 h + R\Omega(\mathbf{x}, t) + \eta. \tag{5.38}$$

The exposure angle Ω measures the amount of particle flux that each point receives. If a surface point has no exposure, it will receive no flux and be subject only to noise and perhaps surface diffusion. In an attempt to model the growth of a surface under both shadowing and reemission effects, a stochastic continuum growth equation has been proposed by Drotar et al. [34], given by

$$\frac{\partial h}{\partial t} = \nu \nabla^2 h - \kappa \nabla^4 h + \sqrt{1 + |\nabla h|^2} \left[\sum_{i=0}^{\infty} s_i F_i(\mathbf{x}, t) \right] + \eta. \tag{5.39}$$

In this equation, s_i is the ith-order sticking coefficient, and $F_i(\mathbf{x}, t)$ is the ith-order flux incident on the surface as a result of reemission, defined recursively as

$$F_{n+1}(\mathbf{x}, t) = (1 - s_n) \int Z(\mathbf{x}, \mathbf{x}', t) F_n(\mathbf{x}', t) \frac{(\hat{\mathbf{n}}_{\mathbf{x}\mathbf{x}'} \cdot \hat{\mathbf{n}})\, P(\hat{\mathbf{n}}_{\mathbf{x}'\mathbf{x}}, \hat{\mathbf{n}}')}{|\mathbf{x} - \mathbf{x}'|^2 + |h(\mathbf{x}) - h(\mathbf{x}')|^2} \mathrm{d}A'. \tag{5.40}$$

In this formula, $\hat{\mathbf{n}}$ is the outward unit normal at the position $\mathbf{x}$, $\hat{\mathbf{n}}'$ is the outward unit normal at the position $\mathbf{x}'$, $\hat{\mathbf{n}}_{\mathbf{x}\mathbf{x}'}$ is a unit normal pointing from $\mathbf{x}$ to $\mathbf{x}'$, and $\hat{\mathbf{n}}_{\mathbf{x}'\mathbf{x}}$ is a unit normal pointing from $\mathbf{x}'$ to $\mathbf{x}$. In addition, $P(\hat{\mathbf{n}}_{\mathbf{x}'\mathbf{x}}, \hat{\mathbf{n}}')$ is the probability per unit solid angle that a particle will be reemitted in the direction of $\hat{\mathbf{n}}_{\mathbf{x}'\mathbf{x}}$, and $Z(\mathbf{x}, \mathbf{x}', t)$ is equal to 1 unless there is no line of sight between the surface heights at positions $\mathbf{x}$ and $\mathbf{x}'$ or $(\hat{\mathbf{n}}_{\mathbf{x}\mathbf{x}'} \cdot \hat{\mathbf{n}})$ is negative, in which case $Z(\mathbf{x}, \mathbf{x}', t)$ equals zero. The term $F_0(\mathbf{x}, t)$ reflects properties of the initial incident particle flux similar to the exposure angle Ω in (5.38), and thus where the shadowing effect is modeled in the growth equation. The remaining terms in the equation are reminiscent of the Edwards–Wilkinson equation, Mullins diffusion equation, and KPZ equation to model growth along the surface normal.

A difficulty arises in computing the $F_i(\mathbf{x}, t)$ terms in the growth equation because, in general, they depend on the entire surface morphology. It is therefore not evident how reemission and shadowing affect the surface growth directly from the continuum equation, and it must be numerically integrated to obtain tangible results. As this nonlocal growth equation is very complex, it is often more straightforward to work with discrete Monte Carlo simulations when modeling surfaces under shadowing and reemission effects to obtain more quantitative results. However, we do discuss an interesting limiting case of this model that can be described analytically, the case where the sticking coefficient is small on all orders [36]. If this is the case, we can represent the sum in (5.39) as

$$\sum_{i=0}^{\infty} s_i F_i(\mathbf{x}, t) = s F(\mathbf{x}, t),$$

where

$$F(\mathbf{x}, t) = F_0(\mathbf{x}, t) + \frac{1 - s}{\pi} \int Z(\mathbf{x}, \mathbf{x}', t) F(\mathbf{x}', t) \frac{(\hat{\mathbf{n}}_{\mathbf{x}\mathbf{x}'} \cdot \hat{\mathbf{n}})\, (\hat{\mathbf{n}}_{\mathbf{x}'\mathbf{x}} \cdot \hat{\mathbf{n}}')}{|\mathbf{x} - \mathbf{x}'|^2 + |h(\mathbf{x}) - h(\mathbf{x}')|^2} \mathrm{d}A'. \tag{5.41}$$

To obtain this form, thermal reemission is assumed, which gives $P(\hat{\mathbf{n}}_{\mathbf{x}'\mathbf{x}}, \hat{\mathbf{n}}') = (\hat{\mathbf{n}}_{\mathbf{x}'\mathbf{x}} \cdot \hat{\mathbf{n}}') / \pi$. The zeroth-order flux $F_0(\mathbf{x}, t)$ is simply the amount of "sky" visible from the position $\mathbf{x}$ on the surface weighted with the incident flux distribution, which can be expressed as

$$F_0(\mathbf{x},t) = \int_{sky} (\mathbf{J}(\theta,\phi)\cdot\hat{\mathbf{n}})\,\mathrm{d}\Omega,$$

where $\mathbf{J}(\theta,\phi)$ is the flux arriving at the surface in the direction of the spherical angles θ and ϕ, and "sky" denotes the limits of integration on θ and ϕ at each point $\mathbf{x}$ where the incident flux is not blocked out by other surface features. If we assume a uniform flux distribution, then $\mathbf{J}(\theta,\phi) = \hat{\mathbf{n}}_{\mathbf{x}\mathbf{x}'}/\pi$. In the limit of small s, (5.41) becomes

$$F(\mathbf{x},t) = \int_{sky} \frac{(\hat{\mathbf{n}}_{\mathbf{x}\mathbf{x}'}\cdot\hat{\mathbf{n}})}{\pi}\mathrm{d}\Omega + \frac{1}{\pi}\int Z(\mathbf{x},\mathbf{x}',t)F(\mathbf{x}',t)\frac{(\hat{\mathbf{n}}_{\mathbf{x}\mathbf{x}'}\cdot\hat{\mathbf{n}})\,(\hat{\mathbf{n}}_{\mathbf{x}'\mathbf{x}}\cdot\hat{\mathbf{n}}')}{\rho^2}\mathrm{d}A', \tag{5.42}$$

where $\rho^2 = |\mathbf{x}-\mathbf{x}'|^2 + |h(\mathbf{x}) - h(\mathbf{x}')|^2$. However, recall that the differential solid angle $\mathrm{d}\Omega$ is defined as

$$\mathrm{d}\Omega = \frac{\mathrm{d}A'}{\rho^2}.$$

The term $Z(\mathbf{x},\mathbf{x}',t)\,(\hat{\mathbf{n}}_{\mathbf{x}'\mathbf{x}}\cdot\hat{\mathbf{n}}')$ restricts the integral to surface points that are not shadowed from the point $\mathbf{x}$, and thus the differential

$$\mathrm{d}\Omega = Z(\mathbf{x},\mathbf{x}',t)\,(\hat{\mathbf{n}}_{\mathbf{x}'\mathbf{x}}\cdot\hat{\mathbf{n}}')\,\frac{\mathrm{d}A'}{\rho^2}$$

provides an angular integration over the surface points that are not shadowed from the surface point $\mathbf{x}$, and allows the integral in (5.42) to be written as

$$F(\mathbf{x},t) = \int_{sky} \frac{(\hat{\mathbf{n}}_{\mathbf{x}\mathbf{x}'}\cdot\hat{\mathbf{n}})}{\pi}\mathrm{d}\Omega + \int_{surface} \frac{F(\mathbf{x}',t)\,(\hat{\mathbf{n}}_{\mathbf{x}\mathbf{x}'}\cdot\hat{\mathbf{n}})}{\pi}\mathrm{d}\Omega.$$

One clear solution of this equation is $F(\mathbf{x},t) = 1$, which implies that the flux is uniform for all orders of reemission when the sticking coefficient is small on all orders. If we use this result in (5.39), we recover the KPZ equation under the approximation of a small surface slope. One example of this behavior is under chemical vapor deposition where, depending on the materials used and the deposition temperature, sticking coefficients can be quite small. From experimental measurements of CVD SiO_2 on Si(100) at different temperatures [116], there is evidence that, as the temperature increases, the growth approaches KPZ dynamics as, at higher temperatures, the sticking coefficient becomes smaller and the preceding argument for small sticking coefficients is valid.

5.3 Numerical Integration Techniques

Due to the complexity of many of these continuum growth models, it is helpful to review the key concepts needed to numerically integrate these continuum equations. In this context, we wish to find the solution to the equation

$$\frac{\partial h(\mathbf{x},t)}{\partial t} = \Phi(\mathbf{x},\{h\},t) + \eta, \tag{5.43}$$

where η is random noise, often taken to be Gaussian, which satisfies the properties

$$\langle \eta(\mathbf{x},t) \rangle = 0 \quad \text{and} \quad \langle \eta(\mathbf{x},t)\eta(\mathbf{x}',t') \rangle = 2D\delta^d(\mathbf{x}-\mathbf{x}')\delta(t-t'). \tag{5.44}$$

There are numerous techniques available to numerically solve partial differential equations such as (5.43), however, for the purposes of thin film growth modeling, sophisticated methods are not required to obtain tangible results.

5.3.1 Euler's Method

The most commonly used method to solve these equations, as well as the most physically intuitive method, is Euler's method. The derivative in (5.43) can be approximated as

$$\frac{\partial h(\mathbf{x},t)}{\partial t} \approx \frac{h(\mathbf{x},t+\Delta t) - h(\mathbf{x},t)}{\Delta t} \tag{5.45}$$

for small Δt. Substituting this expression back into the original equation gives

$$h(\mathbf{x},t+\Delta t) \approx h(\mathbf{x},t) + \Delta t \left[\Phi(\mathbf{x},\{h\},t) + \eta\right]. \tag{5.46}$$

This expression is the algorithm for finding an approximate solution to the equation. Take the surface at time t, and compute how the surface will change in the time interval Δt due to the particle flux and growth effects contained in Φ and the random noise η, which are added to the surface at time t to evolve the surface to the time $t + \Delta t$. If Δt is chosen small enough, the algorithm should provide a reasonable estimate of the solution.

In practice, the success of this technique relies on the specific equation to be integrated. As with (5.39), computing Φ is computationally expensive, and reducing Δt to improve accuracy results in a significantly less efficient algorithm because Φ must be computed many more times. Also, reducing Δt to a very small value can cause loss of significance errors. As a result, Δt needs to be judiciously chosen so as to make the algorithm most efficient without compromising accuracy.

One must also be careful with the implementation of the noise in a numerical algorithm such as Euler's method. For example, suppose η were chosen such that any point on the surface would experience, on average, an RMS deviation of one lattice unit per unit time due to noise. Choosing the standard deviation of η to equal one lattice unit with the time interval Δt equal to one unit of time would give the correct noise strength. Now, suppose one reduced Δt by a factor of ten with the same noise and repeated the integration. Due to the nature of the algorithm, each point would experience an RMS deviation of one lattice unit per iteration, and reducing Δt by a factor of ten would create ten iterations of RMS deviations of one lattice unit per unit time. In other words, to have the condition $\langle \eta(\mathbf{x},t)\eta(\mathbf{x'},t')\rangle = 2D\delta^d(\mathbf{x}-\mathbf{x'})\delta(t-t')$ be consistent for all choices of Δt, we must set $\eta \to \eta' = \eta(\Delta t)^{-1/2}$ in the numerical algorithm so as to cancel out the effect of the choice of Δt in the algorithm. The quantity represented as noise in (5.46) is $\eta\Delta t$, which has a variance of $2D\Delta t$, as was shown in Sect. 5.1.1. It follows that

$$\begin{aligned}\langle \eta'(\mathbf{x},t)\eta'(\mathbf{x'},t')\rangle &= (\Delta t)^{-1}\langle \eta(\mathbf{x},t)\eta(\mathbf{x'},t')\rangle \\ &= (\Delta t)^{-1}\left(2D\Delta t\delta^d(\mathbf{x}-\mathbf{x'})\delta(t-t')\right) \\ &= 2D\delta^d(\mathbf{x}-\mathbf{x'})\delta(t-t'),\end{aligned} \tag{5.47}$$

which is consistent with (5.44). We must make a similar modification for the discrete lattice. If the lattice spacing for a surface is Δx, then the variance of the noise per unit area should be constant irrespective of the choice for Δx [131]. If the surface is d-dimensional, this implies that in Euler's method, the continuum noise η must be replaced by the discrete noise η' as

$$\eta \to \eta' = \eta\sqrt{\frac{1}{(\Delta t)(\Delta x)^d}}. \tag{5.48}$$

For example, suppose we would like to implement noise from a uniform distribution, as this is most readily available when writing simulations. Often, as is the case in the C++ standard library, we can generate random numbers in the range $[0,1]$, which can be offset to the range $[-0.5,0.5]$, which has a mean of zero as required by (5.44). Let us denote this distribution by X. The variance of this distribution is

$$\int_{-1/2}^{1/2} x^2 \mathrm{d}x = \frac{1}{12},$$

but we need this distribution to have a variance of $2D$ to satisfy (5.46). It follows that we should use the following noise in (5.46) with random numbers from the distribution X,

$$\eta' = \sqrt{\frac{24D}{(\Delta t)(\Delta x)^d}}X. \tag{5.49}$$

If a Gaussian distribution with mean 0 and variance 1 is used instead of a uniform distribution, the factor of $24D$ becomes $2D$. Because the function Φ does not obey such a variance condition, no such modification in the numerical algorithm is required for the function Φ.

The uniform random number generator included with the C++ standard library, which can be called through the function `rand()`, returns a random integer in the interval 0 to `RAND_MAX`, where the value of `RAND_MAX` depends on the compiler used, but can be as small as 32,768. This random number generator is far from perfect, and depending on the sensitivity of an algorithm to the quality of the random numbers, this generator may be insufficient. Any random number generator implemented on a computer can only have a finite number of random numbers available, and eventually the random numbers will repeat if enough of them are used. Random numbers generated in this fashion are called pseudo-random numbers because an exact sequence of random numbers can be recovered if the generator is seeded similarly. In the example code given in the appendices, the standard C++ random number generator is used, and the results discussed in the following chapters indicate that this random number generator is sufficient for those simple examples. This can be tested by observing the results of a random deposition and measuring the growth exponent β, which should equal $\frac{1}{2}$ if the numbers are random. If a markedly different behavior is observed, it may suggest that the random number generator is not working well enough to give good statistics, and a more robust random number algorithm should be implemented.

5.3.2 Finite Difference Method

Often in continuum growth equations, derivatives of different orders are encountered, and must be numerically estimated to implement a numerical algorithm. For our purposes, a finite difference approximation is the most convenient approximation scheme [145]. These approximations are derived from appropriate Taylor expansions of the functions of interest. For example, to estimate $f'(x)$, one could use the expansions

$$f(x+\Delta x) = f(x) + \Delta x f'(x) + O((\Delta x)^2) \tag{5.50}$$

$$f(x-\Delta x) = f(x) - \Delta x f'(x) + O((\Delta x)^2), \tag{5.51}$$

which would give the approximations

$$f'(x) = \frac{f(x+\Delta x) - f(x)}{\Delta x} + O(\Delta x) \tag{5.52}$$

$$f'(x) = \frac{f(x) - f(x-\Delta x)}{\Delta x} + O(\Delta x). \tag{5.53}$$

These are known as forward difference and backward difference approximations, respectively, by taking as an approximation the first term in the expansion. However, a more accurate method can be obtained from the expansions

$$f(x+\Delta x) = f(x) + \Delta x f'(x) + \frac{(\Delta x)^2}{2} f''(x) + O((\Delta x)^3) \tag{5.54}$$

$$f(x-\Delta x) = f(x) - \Delta x f'(x) + \frac{(\Delta x)^2}{2} f''(x) + O((\Delta x)^3). \tag{5.55}$$

Subtracting these two equations gives the approximation

$$f'(x) = \frac{f(x+\Delta x) - f(x-\Delta x)}{2\Delta x} + O((\Delta x)^2). \tag{5.56}$$

This approximation is accurate to $O((\Delta x)^2)$, however, it requires the value of the function at both $x+\Delta x$ and $x-\Delta x$. This form is known as a central difference. Of use in thin film modeling are the values of $\nabla^2 h(x,y)$, $\nabla^4 h(x,y)$, and $|\nabla h(x,y)|^2$, which can be derived in a similar manner to the first-order derivatives given above. These finite difference approximations are summarized in Table 5.1.

A C++ implementation of Euler's method utilizing these relations is included in App. B to numerically solve the equation

$$\frac{\partial h(x,t)}{\partial t} = \nabla^2 h(x,t) + \eta(x,t), \tag{5.57}$$

in one spatial dimension with cyclic boundary conditions and Gaussian distributed noise. Gaussian noise can be sampled from a uniform distribution by using a Box–Muller transform [14].

5.3.3 Propagation of Errors

Although the expressions for derivatives given in the previous section are theoretically valid as Δx approaches zero, numerically one can observe significant problems if Δx is "too small". Suppose we are numerically computing the derivative of $\sqrt{x}$. The forward difference method gives the approximation

$$f'(x) \approx \frac{\sqrt{x+\Delta x} - \sqrt{x}}{\Delta x}.$$

Suppose we wish to compute $f'(100)$, and we choose $\Delta x = 10^{-6}$. If we simply use the above formula, the algorithm will compute the difference

$$\sqrt{100.000001} - \sqrt{100}.$$

The value of $\sqrt{100.000001} \approx 10 + 5 \times 10^{-8} = 10.00000005$. However, if the computer arithmetic is not sufficiently precise to carry this many significant digits, it will round this number off to 10, and the computer will return

$$\sqrt{100.000001} - \sqrt{100} = 0,$$

instead of

$$\sqrt{100.000001} - \sqrt{100} = 5 \times 10^{-8},$$

which will clearly give the incorrect value for the derivative. Therefore, depending on the precision of the arithmetic used, choosing Δx too small will lead to round-off errors. A similar situation will arise if Δt is chosen too small in (5.46). In general, computing a derivative numerically is an error-prone process, and if possible one should avoid using derivatives in a numerical algorithm by transforming the problem to an integral equation, which may be less prone to these errors. Unfortunately, for thin film growth models, derivatives are abundant in the growth equations, and it is often sufficient to use the finite difference approximations when numerically solving continuum growth models. The reader is urged to consult a reference on numerical computing when implementing such algorithms to avoid other complications that can arise from a discrete solution method [129, 145].

Table 5.1. Summary of finite difference approximations for common expressions in thin film growth modeling. All approximations are accurate to second order. The expression for the biharmonic term is lengthy, note the location of brackets as they may span more than one line.

$$\frac{\partial h(x,y)}{\partial x} \approx (2\Delta x)^{-1}\left[h(x+\Delta x,y) - h(x-\Delta x,y)\right]$$

$$\nabla^2 h(x,y) \approx (\Delta x)^{-2}\left[h(x+\Delta x,y) - 2h(x,y) + h(x-\Delta x,y)\right] + (\Delta y)^{-2}\left[h(x,y+\Delta y) - 2h(x,y) + h(x,y-\Delta y)\right]$$

$$\nabla^4 h(x,y) \approx \begin{cases} (\Delta x)^{-4}\left[h(x+2\Delta x,y) - 4h(x+\Delta x,y) + 6h(x,y) - 4h(x-\Delta x,y) + h(x-2\Delta x,y)\right] \\ +(\Delta y)^{-4}\left[h(x,y+2\Delta y) - 4h(x,y+\Delta y) + 6h(x,y) - 4h(x,y-\Delta y) + h(x,y-2\Delta y)\right] \\ +2(\Delta x)^{-2}(\Delta y)^{-2}\{4h(x,y) - 2\left[h(x+\Delta x,y) + h(x-\Delta x,y) + h(x,y+\Delta y) + h(x,y-\Delta y)\right] \\ +h(x+\Delta x,y+\Delta y) + h(x-\Delta x,y+\Delta y) + h(x+\Delta x,y-\Delta y) + h(x-\Delta x,y-\Delta y)\} \end{cases}$$

$$|\nabla h(x,y)|^2 \approx (2\Delta x)^{-2}\left[h(x+\Delta x,y) - h(x-\Delta x,y)\right]^2 + (2\Delta y)^{-2}\left[h(x,y+\Delta y) - h(x,y-\Delta y)\right]^2$$

6

Small World Growth Model

In an effort to model nonlocal effects in a continuous manner, we must consider a growth model that accounts for nonlocal correlations across the entire surface. The continuum growth models for shadowing and reemission discussed in Sect. 5.2 are often cumbersome, and leave significant room for improvement to a more concrete and accurate model. In this chapter, we discuss a new growth model based on small world network dynamics that will serve as an example of how to analyze a continuum growth model.

6.1 Introduction

The concepts that are introduced to describe nonlocal effects are best understood in the context of a network. One of the most fundamental concerns when considering the dynamics of a network is its synchronization. If all the nodes in a network landscape are synchronized, they will complete their task in an efficient manner because there are no delays in waiting for certain nodes to catch up to other nodes. Perhaps the most concrete example of these dynamics occurs in parallel computing, where one has a large number of processors linked together with the goal of using the combined processing power to complete a common task. In this type of computing scheme, synchronization is tremendously important because each processor relies on the results of other processors, and if some processors lag behind, it can slow the entire network. It has been shown [146, 169] that similar dynamics can be applied to systems involving protein behavior, social networks, and airport traffic, which all are based on a networked infrastructure. Therefore, it has been important to understand the dynamics of these networks, and investigate strategies to help synchronize the network at low cost.

The simplest networking scheme is called a regular network [43, 70, 71], where each node is linked with its nearest neighbors, and possibly its next nearest neighbors. A regular network is depicted in Fig. 6.1a, where each square represents a node on the network. Although regular networks are the

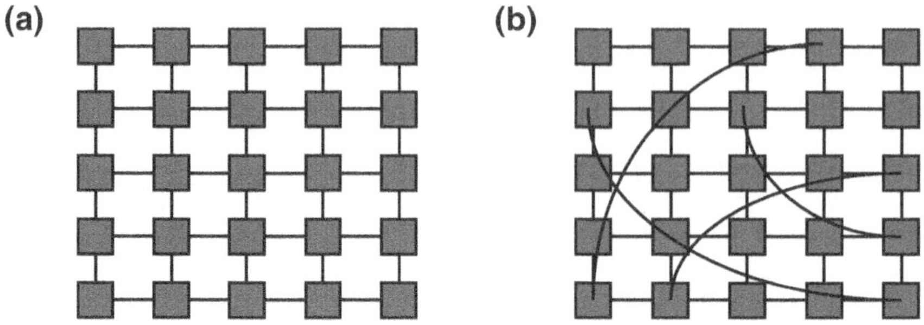

Fig. 6.1. Diagram of (**a**) a regular network and (**b**) a small world network. Each square represents a node in the network, with connections between nodes representing links between nodes. The small world network introduces a relatively small number of long-range links to the regular network.

simplest to implement, they are also susceptible to the problem of desynchronization because information can only travel between adjacent nodes. One strategy that can be used to improve synchronization is simply to connect every node with every other node, so information can travel directly between any two nodes. This strategy is also not desirable, because the number of links would scale as n^2, as opposed to n in the regular network, which may be difficult or even impossible to implement. The trade-off is to construct a regular network with a few long-range links; this type of network is called a small world network [170], and is illustrated in Fig. 6.1b. The concept of a small world network is worth discussing in the context of thin film growth because, as argued later in this chapter, the behavior of nonlocal growth effects may be mapped to small world network dynamics. This provides an interpretation of the growth processes occurring during thin film growth; in each instance where a particle experiences reemission or shadowing, a link is created between two surface heights, much in the same way links are formed in the context of a small world network.

6.2 Growth Equation

The dynamics of a regular network are familiar from the discussion in Chap. 5 because they are governed by the Edwards–Wilkinson (EW) equation,

$$\frac{\partial h_j}{\partial t} = \nu \nabla^2 h_j + \eta. \tag{6.1}$$

In order to describe a network with this equation, one must map the nodes of the network onto a surface, and assign their relative progress a "height" h on the surface. This formalism makes the transition from small world networks to surface growth particularly straightforward. Adjacent nodes on the abstract surface are linked by the Laplacian term in the EW equation, which has the one-dimensional discrete form

$$\nabla^2 h_j = h_{j+1} + h_{j-1} - 2h_j = (h_{j+1} - h_j) + (h_{j-1} - h_j). \quad (6.2)$$

The Laplacian is simply a sum of height differences between adjacent heights. To generalize this concept to links between arbitrary heights on the surface, the following form is used [70],

$$\frac{\partial h_j}{\partial t} = \nu \nabla^2 h_j + \sum_i J_{ij}(h_i - h_j) + \eta, \quad (6.3)$$

where the factor J_{ij} determines the strength of the coupling between two heights on the surface. These factors can be chosen in a number of different ways depending on the types of networks being investigated, and each choice can lead to a different realization of the small world network. The choice of J for thin film growth is motivated by the specific growth effects to be modeled, and different methods for choosing J are discussed in the following sections.

6.3 Reemission

To investigate reemission and its possible connection to small world networks, we consider the "ideal" system with which to study reemission: a deposition with normally incident particle flux and no surface diffusion, where the lack of angular flux prohibits geometrical shadowing from occurring. We aim to find an appropriate method to choose the coupling factor J in (6.3) to model reemission, and one way of doing so would be to guess different forms of J based on the physics of reemission, and compare to existing results. For example, the coupling factor J will clearly depend on the distance between the two coupled heights because a reemitted particle will most likely deposit on a nearby surface height, but it also has a small probability of traveling a far distance after reemission. Another method for investigating the coupling factor J would be to measure correlations created by the reemission effect in another model, and infer the behavior of J from this model. Because reemission is most easily implemented in a discrete Monte Carlo (MC) solid-on-solid model, we choose to investigate the results of a solid-on-solid model that incorporates reemission. The discussion of discrete models is introduced in Chap. 7, but for the present discussion, we only use the results of these models to suggest a reasonable small world model for reemission. Following the discussion of Sect. 4.4.3, a uniform model of reemission is implemented to obtain the results that follow.

The synchronization of the system is reflected by the interface width w, and the behavior of the interface width for different values of the sticking coefficient s_0 is pictured in Fig. 6.2. A unity sticking coefficient implies no reemission, and consequently random growth with $\beta = 0.50$. Smaller values

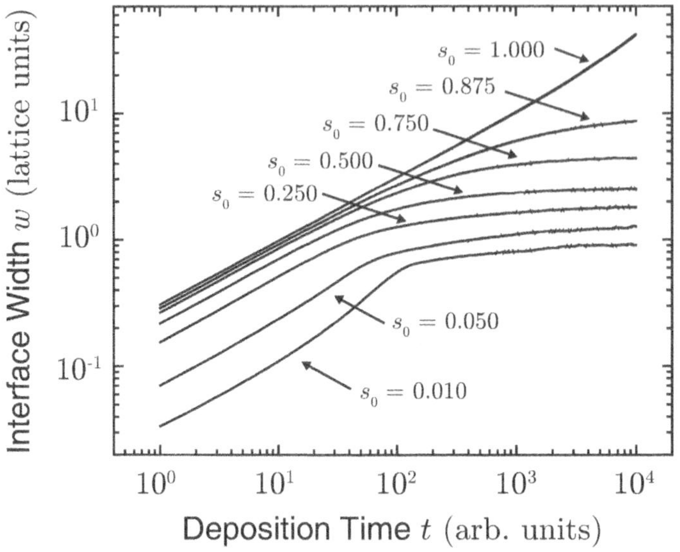

Fig. 6.2. Interface width w versus deposition time t for different values of the sticking coefficient s_0 under a 2+1 dimensional MC model. Stronger reemission (smaller s_0) leads to a smoother surface.

of the sticking coefficient s_0 imply a larger percentage of incident particles that are reemitted, and consequently a smoother surface. At initial times, the surfaces show a roughness evolution characteristic of random growth, but as reemission begins to occur, the surface roughens more slowly.

Due to the discrete nature of the simulation, one can track the trajectories of each particle, and in particular the particles that are reemitted. In Fig. 6.3, a plot of the normalized probability distribution $\mathcal{P}(r)$ for a particle traveling a distance r on the surface after reemission is plotted. The variable $\mathcal{P}$ is used to discriminate between probability $\mathcal{P}$ and the power spectral density function P. Interestingly, the form of this distribution is independent of the sticking coefficient, and takes the form of a power law,

$$\mathcal{P}(r) \sim r^{-\chi}, \tag{6.4}$$

where χ is observed to be $\chi \approx 2.75 \pm 0.10$. If we wish to model this behavior with a continuum equation, the EW equation will not suffice because it does not take these long-range correlations into account in the growth dynamics. However, we should expect the small world growth equation to mimic this growth with an appropriate choice of the coupling factor J. Because $\mathcal{P}(r)$ is the distribution of distances for reemitted particles only, the probability that any incident particle is reemitted a distance r is given by

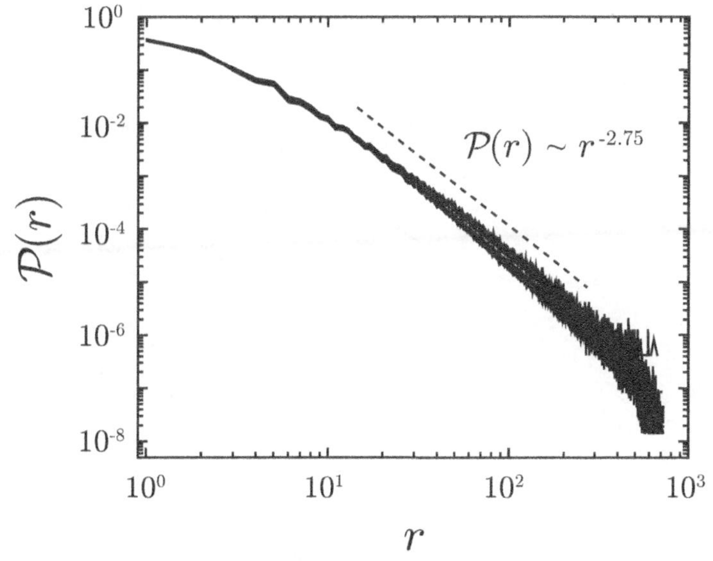

Fig. 6.3. Probability distribution of the distance particles travel upon reemission plotted for sticking coefficients s_0 ranging from 0.0 to 0.9. The distribution is independent of the sticking coefficient, and is a power law with exponent $\chi \approx 2.75 \pm 0.10$.

$$\mathcal{P}(r) = \begin{cases} s_0, & r = 0, \\ (1 - s_0)(\chi - 1)r^{-\chi}, & r \geq 1. \end{cases} \tag{6.5}$$

This distribution tells us how to choose the coupling factor J. Suppose two surface heights are separated by a distance r_0. Then, the probability they are linked is given by $\mathcal{P}(r_0)$. If two surface heights are linked, then the coupling factor between those heights $J_{ij} = 1$, otherwise it is zero. Using this scheme in (6.3), one obtains roughness curves as pictured in Fig. 6.4, which show the same behavior as those from the MC reemission model in Fig. 6.2.

6.4 Shadowing

As opposed to reemission, which tends to reduce the surface roughness, shadowing tends to enhance the surface roughness. Therefore, the method of modeling reemission presented in the previous section does not immediately hold for shadowing because, regardless of the distribution of links, small world links will tend to "synchronize" the surface and reduce the roughness. To model shadowing, consider the physical interpretation of a negative link strength J. Figure 6.5 shows the interpretation of a negative coupling term.

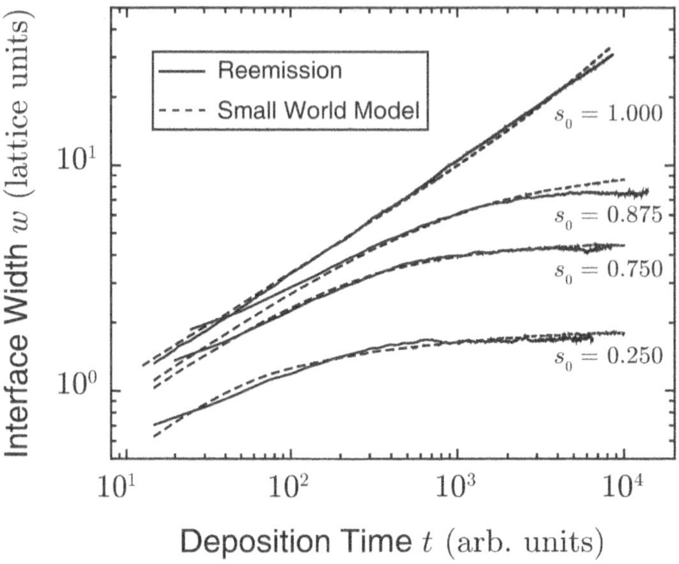

Fig. 6.4. Interface width w versus deposition time t for the small world model with a link distribution chosen according to (6.5), along with roughness curves from Fig. 6.2. The behavior of these curves mimics the roughness behavior found in the MC models for reemission.

It is important to note that (6.3) is given in a frame where the mean height is equal to zero, which is evident from a summation over j in (6.3). First, consider reemission. From a stationary reference frame, in terms of Fig. 6.5a, h_1 does not change, and h_2 increases. However, the introduction of another particle increases the mean height $\overline{h}$. Thus, from a frame that moves with the mean height, h_1 decreases and h_2 increases, which is the prediction of the small world term with $J > 0$ in (6.3),

$$\frac{\partial h_1}{\partial t} \propto (h_2 - h_1) < 0$$
$$\frac{\partial h_2}{\partial t} \propto (h_1 - h_2) > 0.$$

In shadowing, complementary growth dynamics take place. In terms of Fig. 6.5b, we have

$$\frac{\partial h_1}{\partial t} \propto -(h_2 - h_1) > 0$$
$$\frac{\partial h_2}{\partial t} \propto -(h_1 - h_2) < 0,$$

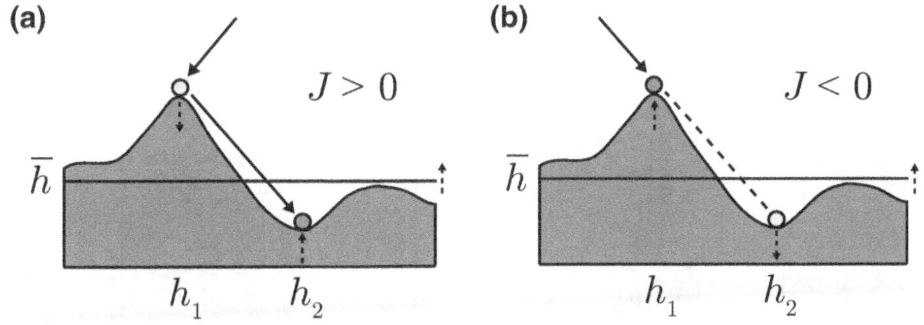

Fig. 6.5. Physical interpretation of the effects of a positive and negative small world coupling term J. In (**a**), if a particle is reemitted from h_1 and deposits on h_2, from a frame where $\overline{h} = 0$, $\Delta h_1 < 0$ and $\Delta h_2 > 0$. This is correctly modeled by a positive small world coupling term ($J > 0$) in (6.3). In (**b**), if h_2 is shadowed by h_1, from a frame where $\overline{h} = 0$, $\Delta h_1 > 0$ and $\Delta h_2 < 0$. This is correctly modeled by a negative small world coupling term ($J < 0$) in (6.3).

which is the result obtained from (6.3) with $J < 0$. The manner in which we choose J in reemission and shadowing is significantly different, but the dynamics of the growth are dictated by the sign of J. Clearly, for shadowing, $J_{ij} \neq 0$ if h_i shadows or is shadowed by h_j.

There is one caveat with this model for shadowing. Suppose we evolve the surface as in Fig. 6.5b with $J = -1$. Then, h_1 will grow without bound because, as h_1 becomes large ($h_1 \gg h_2$),

$$\frac{\partial h_1}{\partial t} \propto -(h_2 - h_1) \approx h_1,$$

which gives an exponential growth for h_1. Therefore, we must normalize the magnitude of J to reflect a constant flux of particles onto the surface. This restraint is imposed to keep the model physically relevant, and has been implemented in other models of shadowing to prevent a similar divergence [177].

To investigate the validity of a model with a negative link strength, let us examine a model where heights are negatively linked if they are separated by a distance less than a prescribed distance λ_0. If so, the coupling factor J is a negative constant, otherwise it is zero. If the previous discussion regarding negative links in a small world model is correct, we should obtain a mounded surface with wavelength λ_0. The PSD of a surface evolved under such a model with $\lambda_0 = 10$ lattice units is pictured in Fig. 6.6a, which shows a mounded surface with wavelength $\lambda = 12.47 \pm 2.5$ lattice units. Varying λ_0 leads to a corresponding change in λ, as is the case in Fig. 6.6b with $\lambda_0 = 20$ lattice units. The measured wavelength λ is slightly larger than λ_0 due in part to the discrete nature of the lattice and the random noise inherent to the model, but the important result is that a small world model with a negative link strength does give a mounded surface, even though this particular model is

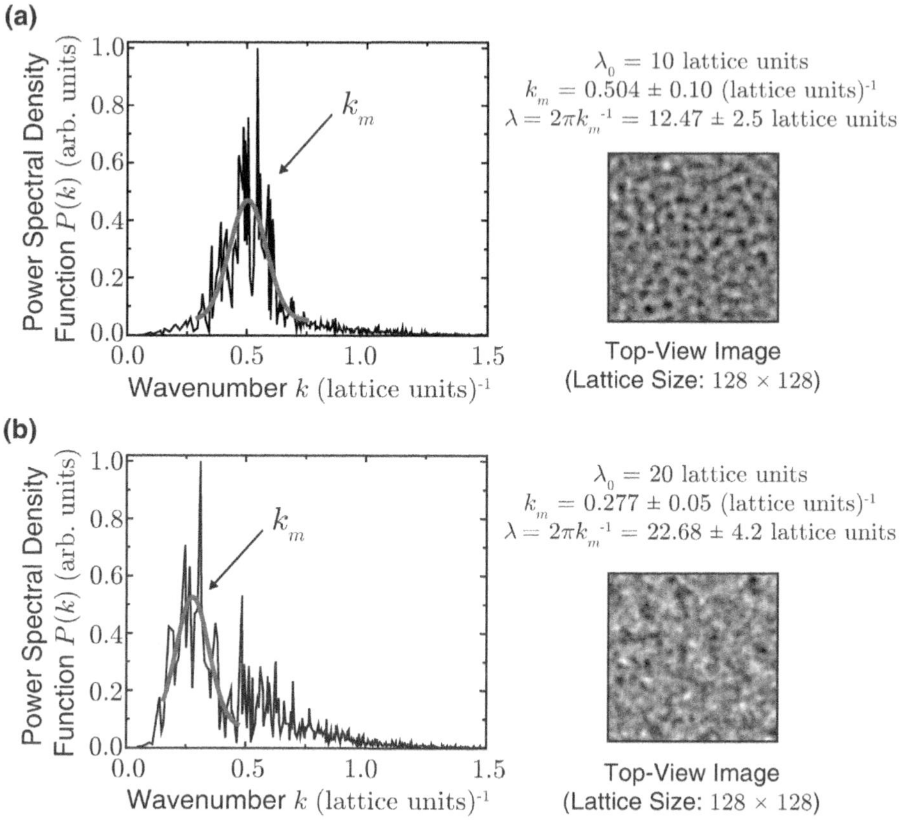

Fig. 6.6. Power spectral density functions for surfaces evolved under negative links, where two heights are linked if they are within (**a**) $\lambda_0 = 10$ lattice units and (**b**) $\lambda_0 = 20$ lattice units of each another. A mounded surface is realized with wavelength λ approximately equal to λ_0.

not physical. The code used to generate these results is included in App. C, which can be generalized to model the other link distributions introduced in this chapter.

To generate a physical shadowing model, we must determine how particles are shadowed by surface heights, which will vary depending on the profile of the incident particle flux. The simplest flux to consider is an oblique flux, where every particle approaches the surface at an angle θ from the surface normal. This allows for a simple determination of shadowing links on the surface, as shown in Fig. 6.7. Two surface points at locations $\mathbf{x}$ and $\mathbf{x}'$ on the surface are linked by shadowing under oblique angle deposition if

$$|\mathbf{x} - \mathbf{x}'| \leq |h(\mathbf{x}) - h(\mathbf{x}')| \tan\theta. \tag{6.6}$$

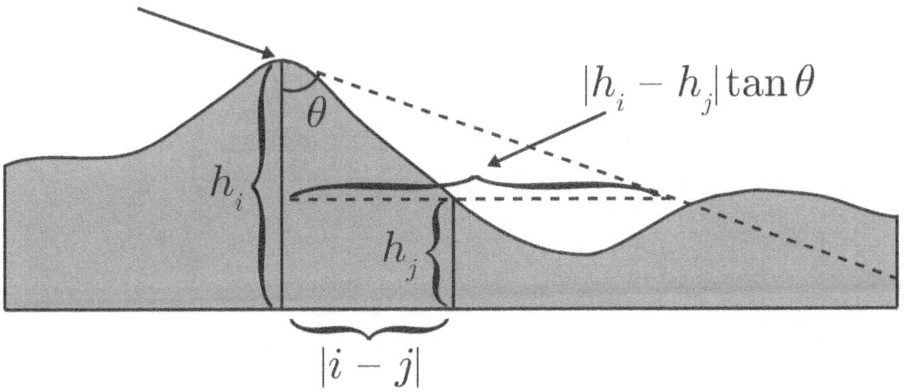

Fig. 6.7. Schematic diagram for determining if two heights are linked under shadowing. Under an oblique flux of angle θ, the heights h_i and h_j are linked by shadowing if $|i - j| \leq |h_i - h_j| \tan\theta$.

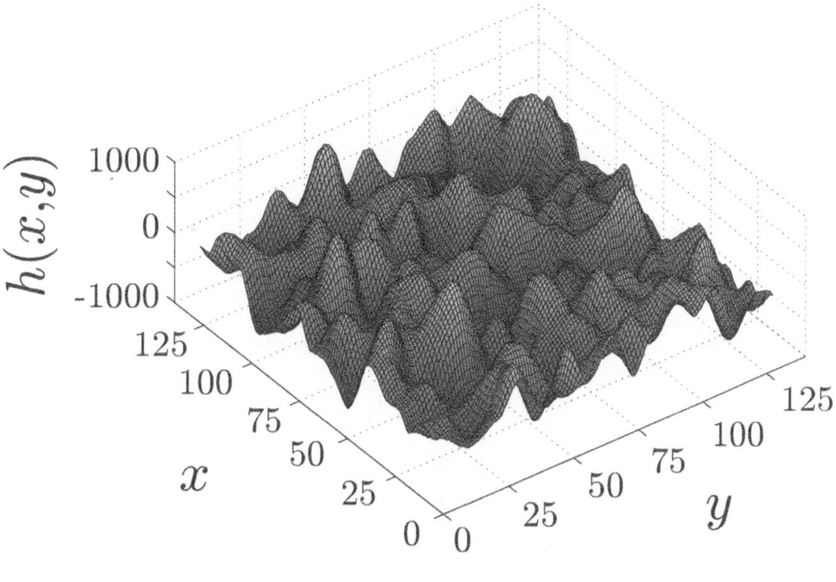

Fig. 6.8. Surface evolved under the small world model with negative links distributed according to (6.6) with $\theta = 85°$.

Using this condition to choose the coupling factor J, with $\theta = 85°$, leads to the surface shown in Fig. 6.8 and the statistics in Fig. 6.9.

The statistics most indicative of shadowing behavior are the exponents β and p, which describe the time evolution of the interface width and wavelength, respectively. Previous work [123] indicates that under strong geometrical shadowing, $\beta = 1$ and $p = 0.50$, which are both within the error of the statistics measured from the small world model. The value of the exponent

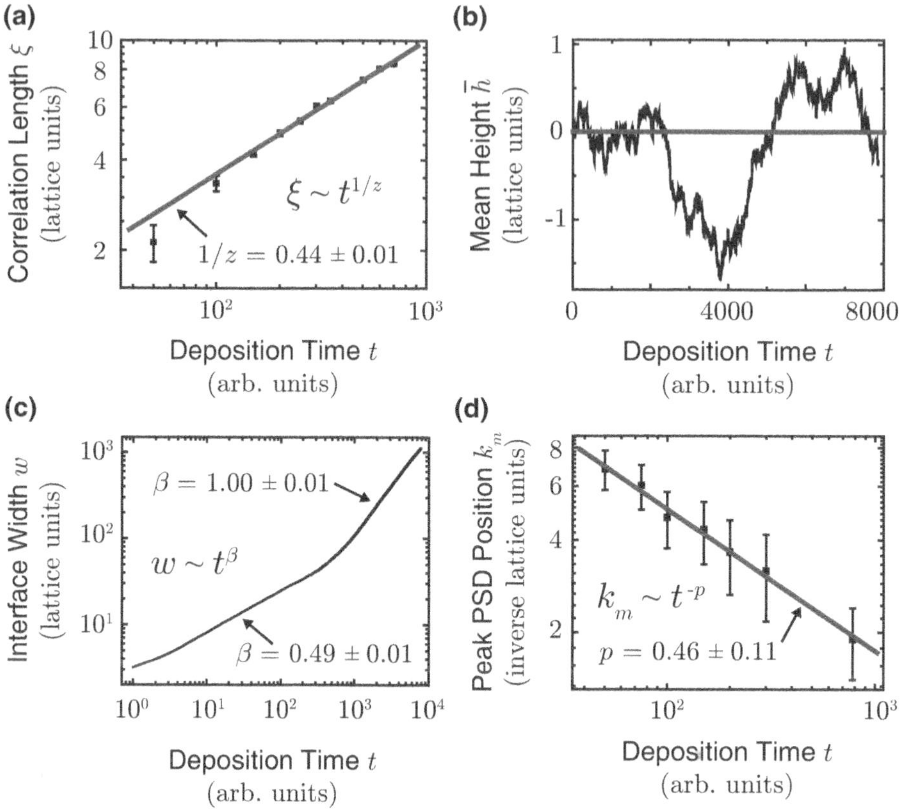

Fig. 6.9. Surface statistics of the surface pictured in Fig. 6.8 including (**a**) the lateral correlation length ξ, (**b**) the mean height $\overline{h}$, (**c**) the interface width w, and (**d**) the PSD peak position k_m. The mean height exhibits a random walk about 0, and all other statistics are consistent with experimental results for mounded surfaces [122].

characterizing the lateral correlation length, $1/z$, is sensitive to local effects such as the strength of diffusion. However, the value of $1/z = 0.44 \pm 0.01$ is certainly reasonable for this type of growth [122].

For a more complicated flux distribution, as is encountered in sputter deposition and chemical vapor deposition, the link structure can be derived by considering the result obtained for oblique angle deposition. For example, the flux distribution in chemical vapor deposition is often modeled with a cosine distribution, where the probability that a particle has a trajectory in the (θ, ϕ) direction behaves as $\cos\theta$. To derive the algorithm for choosing the coupling factor J, we can rewrite the probability that two sites $\mathbf{x}$ and $\mathbf{x'}$ are linked under an oblique flux of angle $\theta = \theta_0$ from (6.6) as

$$\mathcal{P}(\mathbf{x}, \mathbf{x'}) = \Theta\left(|h(\mathbf{x}) - h(\mathbf{x'})| \tan\theta_0 - |\mathbf{x} - \mathbf{x'}|\right), \tag{6.7}$$

where $\Theta(x)$ is the Heaviside function, $\Theta(x) = 0$ for $x < 0$, and $\Theta(x) = 1$ for $x \geq 0$. Now consider an incident particle in chemical vapor deposition. Each particle will experience the same shadowing behavior as in oblique angle deposition, but each particle will have a different impingement angle θ. Thus, the probability that two sites will be linked is the probability that a particle will have a trajectory in the direction of θ, multiplied by the probability that such a particle will be shadowed, which is given by (6.7), and then integrating over all angles. This can be written as

$$\mathcal{P}(\mathbf{x}, \mathbf{x}') = \int \left(\frac{\mathrm{d}\mathcal{P}}{\mathrm{d}\Omega}\right) \Theta\left(|h(\mathbf{x}) - h(\mathbf{x}')| \tan\theta - |\mathbf{x} - \mathbf{x}'|\right) \mathrm{d}\Omega. \tag{6.8}$$

The Heaviside function in the integrand simply serves to limit the domain of integration over θ to an interval $\theta \in [\theta_c, \pi/2]$, where the critical angle θ_c is defined as

$$\theta_c = \tan^{-1}\left(\frac{|\mathbf{x} - \mathbf{x}'|}{|h(\mathbf{x}) - h(\mathbf{x}')|}\right). \tag{6.9}$$

Using the cosine distribution to model chemical vapor deposition, $\mathrm{d}\mathcal{P}/\mathrm{d}\Omega = \cos\theta/\pi$, this integral becomes

$$\mathcal{P}(\mathbf{x}, \mathbf{x}') = \int_0^{2\pi} \int_{\theta_c}^{\pi/2} \frac{\cos\theta}{\pi} \sin\theta \mathrm{d}\theta \mathrm{d}\phi = \cos^2\theta_c = \frac{|h(\mathbf{x}) - h(\mathbf{x}')|^2}{|\mathbf{x} - \mathbf{x}'|^2 + |h(\mathbf{x}) - h(\mathbf{x}')|^2}. \tag{6.10}$$

The probability that two surface heights are linked by shadowing in chemical vapor deposition is given by (6.10), which can be used to find the coupling factor J for each pair of heights. Note that (6.8) reduces to the result obtained for oblique angle deposition with $\mathrm{d}\mathcal{P}/\mathrm{d}\Omega = \delta(\theta - \theta_0)/(2\pi \sin\theta)$, as

$$\begin{aligned}
\mathcal{P}(\mathbf{x}, \mathbf{x}') &= \int_0^{2\pi} \int_{\theta_c}^{\pi/2} \frac{\delta(\theta - \theta_0)}{2\pi \sin\theta} \sin\theta \mathrm{d}\theta \mathrm{d}\phi \\
&= \int_{\theta_c}^{\pi/2} \delta(\theta - \theta_0) \mathrm{d}\theta \\
&= \Theta(\theta_0 - \theta_c) \\
&= \Theta\left(\theta_0 - \tan^{-1}\frac{|\mathbf{x} - \mathbf{x}'|}{|h(\mathbf{x}) - h(\mathbf{x}')}\right) \\
&= \Theta\left(|h(\mathbf{x}) - h(\mathbf{x}')| \tan\theta_0 - |\mathbf{x} - \mathbf{x}'|\right).
\end{aligned} \tag{6.11}$$

Although we have argued that nonlocal effects can be modeled with a small world network, there is still much to learn from further investigations of this model. For instance, the models discussed in this chapter are under the condition of either strong reemission or strong shadowing, and not a combination of the two growth effects. Studies have been carried out on the nature of the competition between these growth effects [122, 123], and this topic is discussed further in Sect. 8.2.2. However, it is not clear if simply adding together the positive and negative links used to model reemission and shadowing would give the correct crossover behavior, especially because these effects are nonlocal. In addition, it would be interesting to investigate other potential applications of a negatively linked network, even though such a concept in traditional small world networks would be counterproductive as negative links tend to desynchronize the network.

Part III

Discrete Surface Growth Models

7

Monte Carlo Simulations

We begin the discussion of discrete models in thin film growth with Monte Carlo (MC) modeling methods. In general, MC methods rely on introducing a stimulus to a system in a somewhat random fashion, with the aim of discerning the general behavior of the system by averaging over the random process. This method is often used when more concrete numerical methods are unavailable or impractical.

7.1 Monte Carlo Integration

To introduce Monte Carlo methods for our purposes, it is simplest to describe an algorithm that embodies the concepts of a MC model, and later apply those concepts to thin film growth models. To this end, we first introduce a MC model for computing the value of a definite integral that involves a function of high dimension [86]. Consider the integral

$$I = \int_V f(\mathbf{x}) \mathrm{d}\mathbf{x}. \tag{7.1}$$

If the number of grid points in one dimension is K, then the number of points used in a traditional numerical evaluation of the integral would scale as K^d, which can easily become numerically intractable if d is large. As an alternative, one can randomly pick N points $\mathbf{X}_m$ in the domain and obtain an estimate of the integral, which amounts to finding the average value of the function over the domain of integration and multiplying this average by the volume of the domain,

$$I \approx V \left(\frac{1}{N} \sum_{m=1}^{N} f(\mathbf{X}_m) \right), \tag{7.2}$$

where V is simply the volume of the domain of integration, and $\mathbf{X}_m$ are the randomly selected points in the domain. Assuming a well-behaved function

$f(\mathbf{x})$, the variance in the estimate will scale as $1/N$ from the central limit theorem. This algorithm captures the essence of MC methods, taking enough "random shots" at the system will eventually reveal its average behavior.

For a more concrete example of how one would implement a MC algorithm, we examine a canonical MC problem, estimating the value of π. Consider a circle of radius 1 centered at the origin of the (x, y)-plane. The area of the circle in the first quadrant is given by the integral

$$I = \int_0^1 \int_0^{\sqrt{1-x^2}} \mathrm{d}y\mathrm{d}x = \int_0^{\pi/2} \int_0^1 r\mathrm{d}r\mathrm{d}\theta = \frac{\pi}{4}. \tag{7.3}$$

Therefore, if we can obtain a numerical estimate of I, we obtain a numerical estimate of $\pi = 4I$. This reduces to the problem stated earlier for a two-dimensional system, and can be easily carried out with ordinary numerical integration techniques. However, to illustrate MC methods, we opt for a MC algorithm to estimate the integral. We can write the integral I as

$$I = \int_0^1 \int_0^1 f(x, y)\mathrm{d}x\mathrm{d}y, \tag{7.4}$$

where the function $f(x, y)$ is defined as

$$f(x, y) = \begin{cases} 1, & x^2 + y^2 \leq 1, \\ 0, & \text{otherwise.} \end{cases} \tag{7.5}$$

To carry out the algorithm, first choose two independent random numbers x and y uniformly in the interval $[0, 1]$. The sum in (7.2) reduces to counting how many of the N ordered pairs (x, y) satisfy $x^2 + y^2 \leq 1$. Because the volume of the domain is 1, calculating the proportion of ordered pairs that satisfy the constraint should converge to $\pi/4$. Multiplying this result by 4 gives an estimate for π. The results of this algorithm are plotted in Fig. 7.1.

It is apparent that increasing the number of trials N improves the estimate of π. In this specific case, we can compute the standard deviation explicitly because, due to the simplicity of the integrand $f(x, y)$ in I, the method is a binomial process with probability of success $p = \pi/4$. The standard deviation of the total number of successes of a binomial process is given by $\sqrt{Np(1-p)}$ [121], and it follows that the standard deviation of the relative number of successes is $\sigma = \sqrt{p(1-p)/N}$. As a binomial distribution approaches a normal distribution for large N, the interval bounded by $\pm 2\sigma$ represents a 95% confidence interval about the mean. This interval is plotted in Fig. 7.1.

The key idea in this example is that the relative error decreases with increasing sample size, ultimately converging to the true value of the expression. In MC simulations for thin film growth, it is implicitly assumed that increasing the number of "random shots" into a system will give a better estimate of the true behavior of the system, although confirming this assumption is much more complicated than in this simple example, and often impossible because the exact solution is unknown.

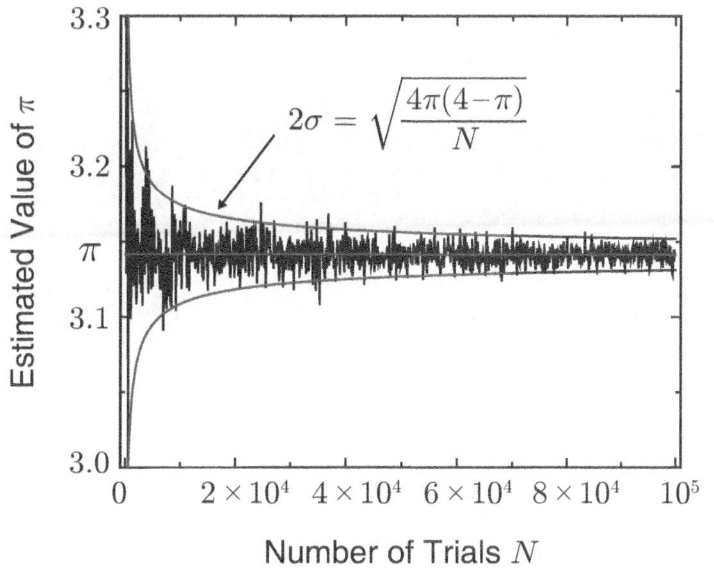

Fig. 7.1. Plot of an estimated value for π obtained from a MC algorithm with N trials. The relative error in the estimate behaves as $1/\sqrt{N}$.

7.2 Structure of Thin Film Growth Models

In practice, MC models in thin film growth evolve under simple rules defined to model particular growth effects. The general execution of such a model is as follows.

- Initialize a lattice on which the deposition will take place. This lattice is usually of two or three dimensions, with a size on the order of 1000 lattice points per dimension. The substrate is taken to be one edge of this lattice.
- Create a particle at a random lattice point, and evolve the particle in time according to a specified trajectory.
- When the particle strikes the substrate, allow it to deposit, or reflect off, depending on deposition parameters.
- Allow particles on the surface to diffuse according to a specified model for diffusion.
- Create a new particle and repeat the deposition process.

Many MC algorithms follow this serial process, but more sophisticated models allow for a parallel execution of the algorithm, which can then be run more efficiently under a parallel computation scheme. Ordinarily, the complexity of a MC algorithm is simple enough that it can be run reasonably

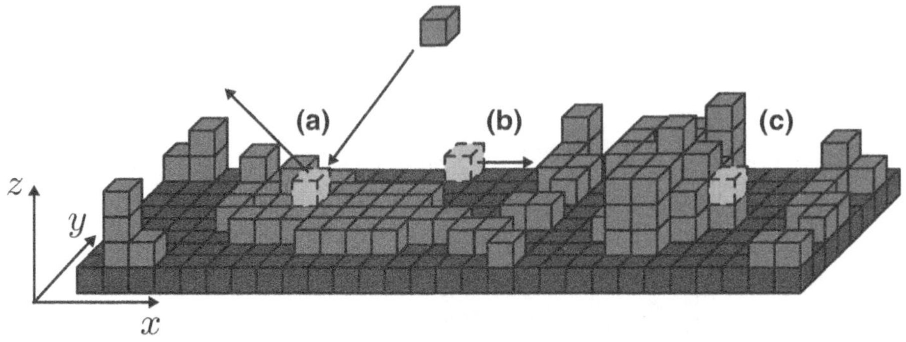

Fig. 7.2. Diagram of the basic processes implemented in Monte Carlo simulations used to model thin film growth. The (**a**) reemission effect, (**b**) surface diffusion, and (**c**) shadowing effect can all be modeled in the simulations [59].

quickly on commercially available computers, and the resources of a supercomputing cluster are not needed. If computation power and resources are an issue, choosing which growth effects to include in a model is often a trade-off between a more physical model and a more efficient model. A graphical representation of some of the growth effects modeled in MC simulations is included in Fig. 7.2 [59].

7.2.1 Particle Modeling

In MC models of thin film growth, each occupied lattice point is taken to represent one particle of the source material. In many growth processes, this is often a single atom, for example, silicon or tungsten in a physical vapor deposition process, but it could also represent a molecule in the context of a chemical vapor deposition process. Monte Carlo models often do not incorporate specifics of the deposition, such as the chemical nature of the deposition flux, but rather leave these effects to be modeled with more empirical parameters that can be easily implemented in the algorithm, such as the activation energy for diffusion or the sticking coefficient that determines the probability that a particle sticks to the substrate when it strikes.

After a particle has been initialized in the context of the algorithm, it must be assigned a trajectory that models a particular growth process. The trajectory can be one of two types: deterministic or stochastic. A deterministic trajectory is one where the particle travels in a straight line according to angles assigned to it when it is initialized, where the randomness is manifested in the selection of such angles and the initial position of the particle. A stochastic trajectory has no assigned direction, and is essentially a random walk or some derivative thereof under which the particle evolves. Which type of trajectory to choose for a particular MC model depends on the type of deposition being modeled. One would expect that, under the conditions of high vacuum normally encountered in physical vapor deposition processes, the mean free path of a particle is much longer than the distance it travels between the source and the substrate, and the assumption that it travels in a straight line is a valid assumption [96]. Models that utilize this assumption are called solid-on-solid or ballistic aggregation models depending on whether overhangs are allowed on the surface. On the other hand, in processes where the deposition pressure is high, diffusion is the primary transport mechanism, and the Brownian motion characteristic of diffusion is best modeled with a random walk. These types of simulations are commonly referred to as diffusion-limited aggregation (DLA) [171], and are often used to model transport phenomena in fluids. The experiments of consideration here are performed under high vacuum, and the deterministic trajectory assumption is used. In addition, periodic boundary conditions are imposed on the lattice, which means that if the trajectory of a particle takes it off the edge of the lattice, it will reappear on the opposite side of the lattice with the same trajectory.

If deterministic trajectories are implemented in the model, we must specify how to choose the initial position and direction of the trajectory. First, we make the assumption that the distribution of particle trajectories is independent of position. In other words, we can choose the initial position of a particle independent of the trajectory because of this uniformity. As such, the initial position of a particle is often randomly chosen in the domain. The specific type of deposition is reflected in the distribution of velocities of the particles. This distribution is normally expressed as the probability $\mathrm{d}\mathcal{P}$ of choosing a trajectory in the direction $\mathrm{d}\Omega$,

$$\frac{\mathrm{d}\mathcal{P}}{\mathrm{d}\Omega} = f(\theta, \phi). \tag{7.6}$$

The simplest particle flux is normally incident flux, where every particle impinges normally onto the surface. In this case, with the positive z-axis pointing in the direction of particle flux, $f(\theta, \phi) \propto \delta(\theta)$, and all particles travel parallel to the positive z-axis. In sputter deposition and chemical vapor deposition, experimental data suggest that the probability of a particle obtaining a trajectory making an angle θ with the z-axis is proportional to $f(\theta, \phi) = \cos\theta$, $\theta \in [0, \pi/2]$ [59], as was shown in Fig. 1.2.

To model these distributions numerically, we must be able to sample an arbitrary distribution from a uniform distribution $[0, 1]$, as this is the distribution available in most computing packages. The mathematical term for this process is "inverse transform sampling," and is described in [27]. Using this method, with a uniformly distributed random variable X and cumulative density function (CDF) F, in order to sample from F, one can sample from $F^{-1}(X)$, where F^{-1} is the inverse CDF of the distribution. For example, if we wish to model chemical vapor deposition, we want to sample from the probability distribution function $f(\theta) = \cos\theta$. The CDF of this distribution is

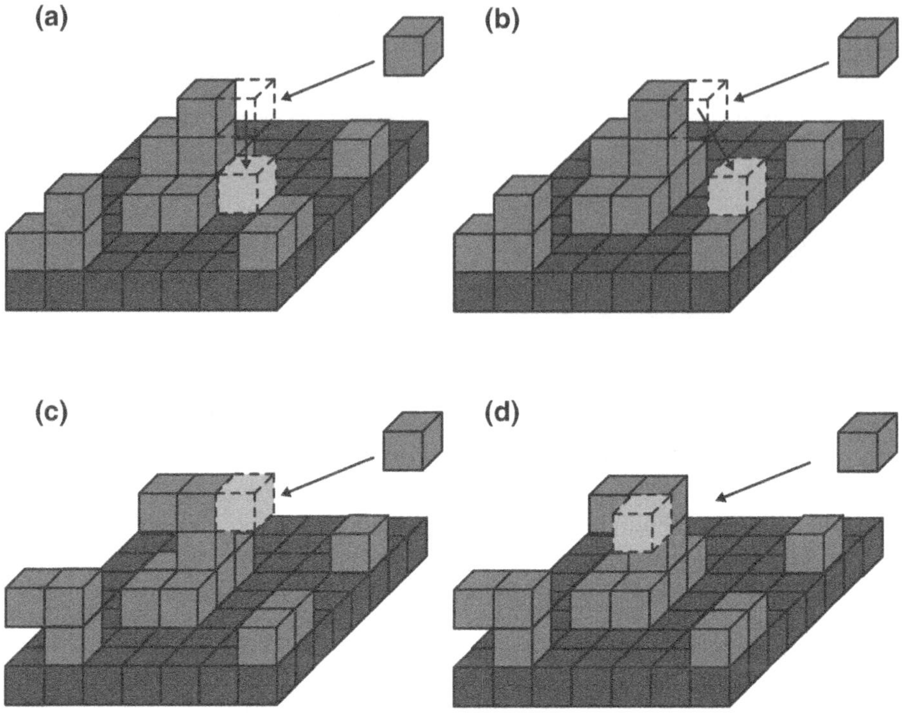

Fig. 7.3. Diagram of different schemes for particle aggregation: (**a**) solid-on-solid aggregation, (**b**) reemission, (**c**) head-on ballistic aggregation, and (**d**) side-sticking ballistic aggregation. Note that in the solid-on-solid model in (**a**), no overhangs are allowed, whereas the ballistic models in (**c**) and (**d**) allow overhangs.

$$F(\theta) = \int_0^\theta \cos\theta' \mathrm{d}\theta' = \sin\theta. \tag{7.7}$$

It follows that the inverse CDF is $F^{-1}(x) = \sin^{-1} x$. Therefore, if we wish to choose a trajectory for a particle while modeling chemical vapor deposition, we choose the angle ϕ from a uniform distribution $[0, 2\pi]$, and the angle θ by selecting a uniform random number in the interval $X \in [0, 1]$ and assigning $\theta = \sin^{-1}(X)$, giving the desired cosine distribution.

7.2.2 Aggregation

Once a particle has collided with the substrate, or any particles previously deposited on the substrate, the model must determine how the aggregate is changed by the addition of a new particle. The most common aggregation schemes are depicted in Fig. 7.3. The simplest way to add a particle to the aggregate is to allow the particle to drop down to the lowest unoccupied height at a certain position on the surface, which is known as solid-on-solid aggregation. This is the easiest to implement because the resultant surface can be described by a single-valued function $h(\mathbf{x})$, as any multiple values of the height at a point $\mathbf{x}$ on the substrate are eliminated. The drawback of this model is that it may not be physical for a particle to simply drop once it hits the surface, although diffusion may bring the particle down to the lowest unoccupied site. The alternative is to include ballistic aggregation, where the particle can attach to any point on the surface. However, for ballistic aggregation, the model must store the lattice in a three-dimensional array as opposed to a two-dimensional array in solid-on-solid aggregation, which can reduce the efficiency of the model. Even so, ballistic models may be more realistic in the sense that particles will tend to remain near the impact point to form aggregates with the possibility of overhangs. All the aggregation schemes in Fig. 7.3 are called on-lattice aggregation models because the aggregation occurs within the constraints of a cubic lattice. Other models, called off-lattice aggregation models, allow particles to aggregate outside the constraints of a lattice, which would be useful if the particles were modeled as spheres instead of cubes because spheres can aggregate at any angle [108].

The dynamics of particle aggregation can be significantly altered if reemission is included in the model. The reemission effect occurs when an incident particle does not stick upon first impact, and can "bounce around" before settling at an appropriate site on the surface. The probability of a particle sticking to the surface on first impact is governed by the sticking coefficient s_0, which gives the probability that a particle will stick when it first strikes the surface. This concept can be generalized to higher-order sticking coefficients (s_n) that describe the probability that a particle will stick after n attempts.

The sticking coefficient is a representative example of the somewhat empirical nature of MC models. Implementing reemission in a MC simulation is trivial, as once the sticking coefficient is defined, a random process is introduced that determines if a given particle will stick or bounce off the surface. The nature of MC models allows for an average over many reemission events, whose effect would be difficult to predict without such an ensemble. However, determining the value of the sticking coefficient from first principles is difficult as it will depend on factors such as particle energy, particle mass, interatomic forces, substrate temperature, and the nature of the particle flux. Thus, implementing a first-principles model that includes reemission would be complex, but with the aid of MC models, we can predict the effects of reemission with relative ease and obtain quantitative predictions to compare with experimental data.

7.2.3 Diffusion

Models for surface diffusion in the literature can vary depending on the specifics of the deposition. A common model for diffusion relies on equilibrium Boltzmann statistics of the particle and substrate at a temperature T. In this model, the diffusing surface atom can jump to a nearby site with a probability proportional to $\exp[-(E_a + n_n E_n)/kT]$ [60], where E_a is the activation energy for diffusion, E_n is the bonding energy with a nearest neighbor, n_n is the number of nearest neighbors, and k stands for the Boltzmann constant. Some models for diffusion also incorporate the effects of next-nearest neighbors, depending on the activation energies and range of attraction. The diffusing particle is also prohibited from making a single jump up to a site where the height change is more than one lattice unit. The particle continues diffusing until it finds a lattice point where $(E_a + n_n E_n)$ becomes large and the diffusion probability becomes small.

A more general model for diffusion can also be implemented that borrows from the Boltzmann model. In this model, after a particle sticks to the aggregate, a particle chosen randomly near the impact point is chosen to diffuse [1, 59] to a nearby position. A particle diffuses if it moves to a site with a larger coordination number than does the present site, where the coordination number is defined by the number of nearest neighbors or next-nearest neighbors at a particular site. This diffusion step is repeated D times per impact. Previous work [179] suggests that a value of $D = 100$ is a reasonable diffusion strength for materials such as silicon deposited by thermal evaporation at room temperature. Another diffusion model similar to this model has any particle on the surface available for diffusion at any time, not just those near a newly deposited particle. Again, when modeling diffusion, more detailed diffusion schemes will be less efficient, and the complexity of the implemented diffusion scheme is up to the discretion of the investigator. If diffusion is a key mechanism in the growth dynamics, it will likely be worth the effort to create a more realistic model for diffusion, but if diffusion is much less important than other growth effects, the simple models discussed in this section should suffice.

8

Solid-on-Solid Models

As was discussed in the previous chapter, solid-on-solid models are ones where no overhangs are allowed on the simulated surface. These models are the simplest to implement because the height profile is a single-valued function of position, and as a result, are the most common discrete models utilized in modeling thin film growth. More complicated models, including ballistic aggregation models, have been of interest as well, but illustrating the formulation and use of solid-on-solid models first gives insight into the advantages and drawbacks of both types of models. For more discussion regarding solid-on-solid models, see [40].

8.1 Local Models

To introduce solid-on-solid models, we examine a simple example of a solid-on-solid model and discuss its various execution steps and results. The model we discuss attempts to model a deposition with normally incident flux that experiences surface diffusion. The C++ implementation of this example is provided in App. D.

The basic execution of the example is as follows. First, a position is chosen randomly above the substrate, and the particle deposits on this position. Then, due to the finite temperature of the substrate, any particle may diffuse with a probability dependent on the activation energy for diffusion E_a, nearest neighbor bond strength E_n, as well as the temperature of the substrate T_s, through the Boltzmann factor, $\exp\left[-(E_a + n_n E_n)/(kT_s)\right]$, where k is the Boltzmann constant and n_n is the number of nearest neighbors for the diffusing particle. If a particle is active for diffusion, it may diffuse to any adjacent surface point with a lower height, and continue diffusing until the probability for diffusion becomes low, and the particle is chosen to stop diffusion. This process is then repeated a determined number of times, given by the variable `jumps` (also denoted as D/F), which represents the number of

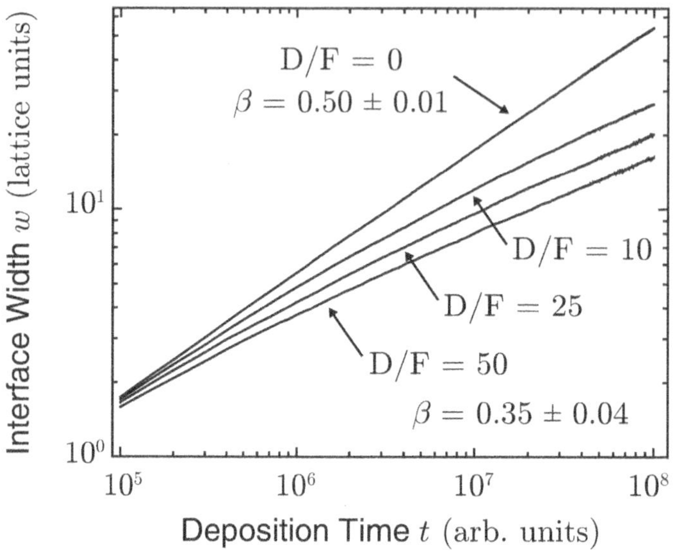

Fig. 8.1. Results of the example solid-on-solid diffusion simulation given in App. D. The variable D/F represents the strength of surface diffusion. For no diffusion, a random deposition model is realized, with the value of β decreasing as diffusion becomes stronger.

particles available for diffusion per unit incident flux [1, 59]. After the diffusion has been carried out, another particle is added to the system in a similar manner. During the simulation, the mean height and surface roughness are output into a file named `stats.txt`, and the surface profile at the end of the simulation is saved in an array to the file `surface.txt`. The code also outputs the autocorrelation function for the surface at intervals throughout the simulation.

For simplicity, the simulation provided in App. D is in 1+1 dimensions, but we also discuss the results of generalizing the model to 2+1 dimensions. The simplest choice of the deposition parameters is to set `jumps = 0`, which deactivates the surface diffusion mechanism. This set of parameters would be analogous to the continuum random deposition model discussed in Chap. 5. The results of this simulation on a lattice of size $N = 32,768$ lattice units, along with results from simulations including surface diffusion, are given in Fig. 8.1. For no diffusion, the roughness shows a strong power-law behavior with $\beta = 0.50$, as was predicted with the continuum random deposition model. Other surface statistics such as the height–height correlation function are not important because there is no lateral correlation between surface heights. This run can be regarded as a check on the simulation to observe if the code is

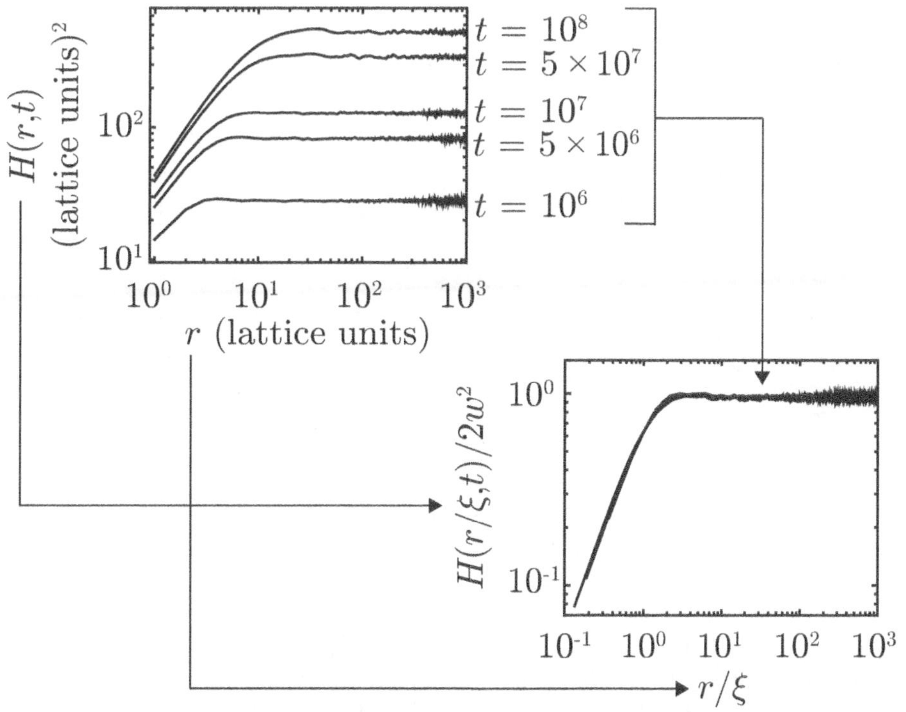

Fig. 8.2. Plot of the height–height correlation function for the solid-on-solid diffusion simulation with $D/F = 50$ at various simulation times. Scaling the horizontal axis by the correlation length ξ and the vertical axis by $2w^2$ collapses all curves onto one, as predicted by dynamic scaling.

working properly, and if the random number generator is giving reasonable random numbers.

The surface roughening behavior becomes more interesting when diffusion is active. If we set the activation energies $E_a = 0.08$ eV and $E_n = 0.05$ eV and vary `jumps` from 10 to 50, we obtain the interface width behavior pictured in Fig. 8.1. The inclusion of diffusion reduces the absolute value of the interface width, but also reduces the value of the growth exponent β, indicating that the interface width is growing more slowly than in the random deposition model. The value for β is reduced from $\beta = 0.50$ in the model without diffusion to approximately $\beta = 0.35$ in the model with the strongest diffusion. This is consistent with the prediction of the Mullins diffusion continuum model presented in Sect. 5.1.4 with $d = 1$.

The measurement of β is performed by fitting a line to the interface width on a log–log scale, and measuring the slope of the line as β. However, the value of β measured will depend on the range of deposition times used for the fitting. For example, in the interface width curves in Fig. 8.1, there is a

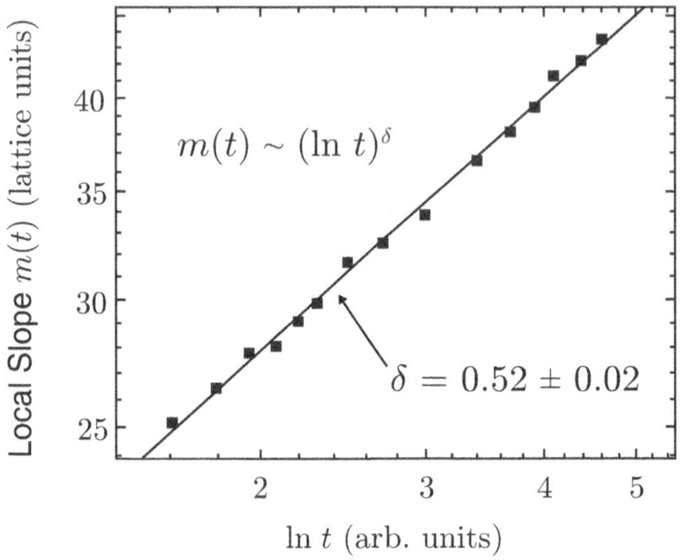

Fig. 8.3. Plot of local slope m versus the logarithm of the deposition time $\ln t$ on a log–log plot for the solid-on-solid diffusion simulation. The slope of the line, $\delta = 0.52 \pm 0.02$, implies that the local slope behaves as $m(t) \sim (\ln t)^{0.52 \pm 0.02}$.

crossover between the roughness evolution from a random deposition, which has $\beta = 1/2$, and the regime where surface diffusion dominates, which gives $\beta < 1/2$. The crossover is gradual, and it is up to one's own judgment from what range of deposition times the value for β will be extracted. This naturally leads to a measurement error for β, and it is wise to fit to many different ranges of time and take an average to report as β. In these simulations, larger values of D/F lead to slightly smaller β values, as $\beta \approx 0.38$ for $D/F = 10$, whereas $\beta \approx 0.32$ for $D/F = 50$, however, these values are within measurement error of each other, and an overall value of $\beta = 0.35 \pm 0.04$ is reported in Fig. 8.1.

For further analysis, we turn to the specific simulation with `jumps = 50`, and examine the behavior of the height–height correlation function. A plot of the height–height correlation function at different deposition times is included in Fig. 8.2. These curves exhibit time-dependent scaling, as rescaling the horizontal axis by the correlation length ξ and the vertical axis by the value $2w^2$ collapses all curves onto one time-independent curve, as predicted by dynamic scaling. Note that, for small r, the unscaled height–height correlation functions in Fig. 8.2 do not overlap, which suggests that the local slope m is not stationary. This behavior was discussed in the context of the Mullins diffusion model in Sect. 5.1.4, in particular Fig. 5.3. We can examine the time-dependence of the local slope m using (3.21). From the small r

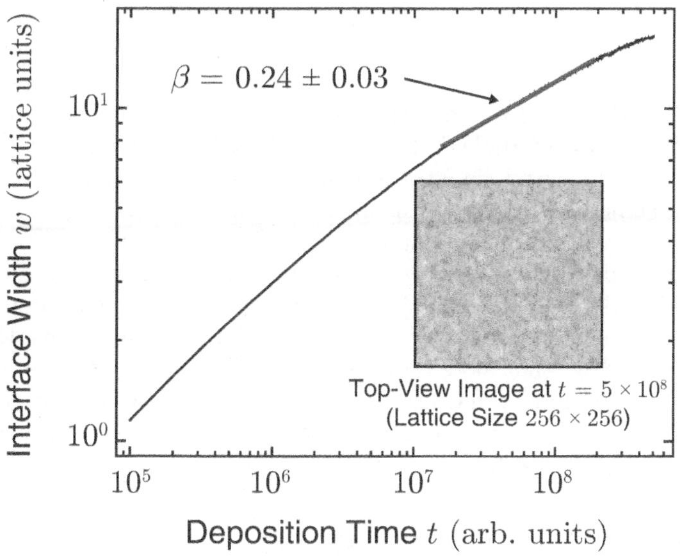

Fig. 8.4. Interface width w versus deposition time t for the solid-on-solid diffusion simulation in 2+1 dimensions, with $\beta = 0.24 \pm 0.03$. The inset is a top-view image of the surface profile at the end of the simulation.

behavior of the height–height correlation function, $\alpha \approx 0.5$, which implies that the local slope can be approximated from the discrete data as $m \sim H(1)$. A plot of the local slope m versus the logarithm of the deposition time $\ln t$ on a log–log plot is included in Fig. 8.3, which implies that the local slope behaves as $m(t) \sim (\ln t)^{0.52}$, similar to the behavior observed experimentally in depositions dominated by surface diffusion [91], which found $m(t) \sim \sqrt{\ln t}$.

A generalization of this model to 2+1 dimensions is straightforward from the simulation code in App. D, but unfortunately the statistics converge much more slowly in 2+1 dimensions as compared to 1+1 dimensions, which makes the analysis more difficult. Therefore, the preceding discussion was carried out in 1+1 dimensions, as the analysis is similar in both cases. We include a graph of the interface width evolution in 2+1 dimensions with $D/F = 100$ on a 256×256 lattice in Fig. 8.4. The growth exponent β is approximately 0.24 ± 0.03, consistent with the prediction of the Mullins diffusion model with $d = 2$, which gives $\beta = \frac{1}{4}$.

8.2 Nonlocal Models

When considering nonlocal growth effects, the simplicity of solid-on-solid models becomes particularly useful because, as was discussed in Chap. 5, analytical models for shadowing and reemission are very complex. One of the first discrete models for shadowing was the needle model [107], which is sometimes called the grass model for the resemblance of the model to the competition of blades of grass for sunlight. In the simplest version of this model, each lattice point on a one-dimensional surface represents a column, and a column grows if it is not shadowed by any other column, where shadowing is defined through an oblique flux of angle θ. As a result, there is a competition between surface heights as shadowed columns die out due to other columns becoming tall. There are some limitations of this model, the first being a neglect for the lateral growth of each individual column as only vertical growth is taken into account. As a result, each column is negligibly thin, which presents a problem when the model is generalized to 2+1 dimensions. If each column has no width, shadowing is ill defined in a 2+1 dimensional setting. Nevertheless, the ideas presented by this model are useful in understanding the behavior of more complicated models that are presented in this section. Most notably, the inclusion of lateral growth can lead to two length scales simultaneously defined on the surface, which may lead to a breakdown of dynamic scaling.

8.2.1 Breakdown of Dynamic Scaling

As was shown in Sect. 4.3, when the lateral correlation length ξ and wavelength λ of a mounded surface evolve at a different rate, the PSD of the surface profile does not scale in time, evidence that the dynamical scaling behavior of the surface has broken down. In this section, we aim to measure the exponents p and $1/z$, which measure the time evolution of the wavelength and lateral correlation length, respectively, in order to test the hypothesis that, under the shadowing effect, the dynamical scaling behavior of a mounded surface breaks down [123].

We begin by measuring p and $1/z$ from a MC model. The simulations in this section are 2+1 dimensional solid-on-solid models with an angular incident flux distribution of $\cos\theta$, where θ is defined with respect to the surface normal. The results of all MC simulations are summarized in Table 8.1. Wavelength selection is indicated by measuring a value for p, and is clear in the simulations where reemission is weak. Figure 8.5 shows simulated surface profiles with $s_0 = 1$ and $D/F = 100$, in the regime of strong wavelength selection. Figure 8.6 contains a plot of the wavelength λ as a function of time for this simulation, where the wavelength exponent $p = 0.49 \pm 0.02$. From Table 8.1, when the sticking coefficient s_0 is reduced in the simulations, the value of the wavelength exponent remains relatively constant at $p \approx 0.5$. However, once the sticking coefficient is sufficiently small ($s_0 < 0.5$), the reemission effect is strong enough to redistribute a significant amount of particle flux to otherwise shadowed surface heights, which effectively cancels the shadowing effect and eliminates wavelength selection. Also, from Table 8.1, varying the strength of surface diffusion (D/F) does not have a significant effect on the wavelength exponent p. Because diffusion is a local growth effect, it is not as

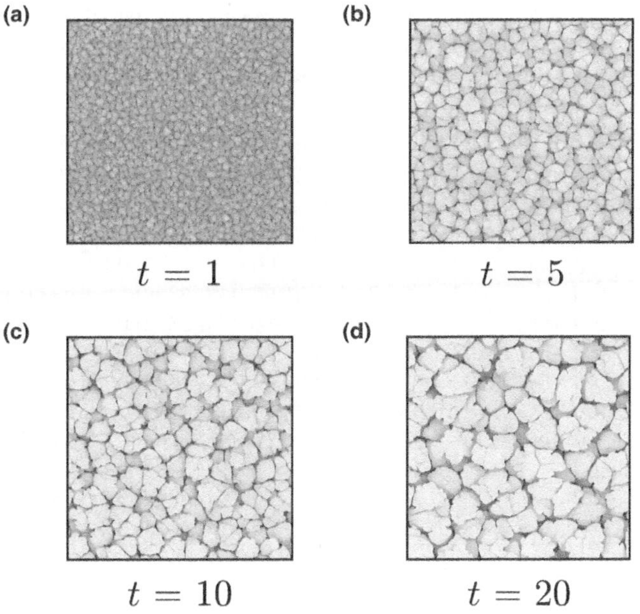

Fig. 8.5. Simulated surface profiles with sticking coefficient $s_0 = 1$ and $D/F = 100$. The deposition time t is defined such that one time step corresponds to an average of 50 deposited particles per lattice point. The size of each image is 512×512 lattice units [122].

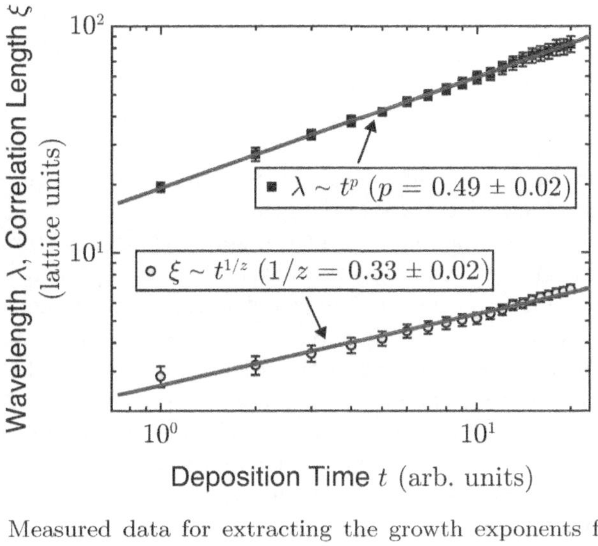

Fig. 8.6. Measured data for extracting the growth exponents for the simulation with sticking coefficient $s_0 = 1$ and $D/F = 100$ (see Fig. 8.5). The extracted values for the exponents are $p = 0.49 \pm 0.02$ and $1/z = 0.33 \pm 0.02$ [122].

Table 8.1. Results of MC simulations under a cosine flux distribution for different values of the sticking coefficient s_0 and strength of surface diffusion D/F. Wavelength selection is only observed for larger values of the sticking coefficient ($s_0 \geq 0.5$).

s_0	D/F	p	β	α	$1/z$	β/α
1.000	0	0.51±0.02	1.00±0.01	0.67±0.03	0.41±0.01	1.49±0.07
1.000	20	0.50±0.02	1.00±0.01	0.59±0.01	0.40±0.04	1.69±0.03
1.000	100	0.49±0.02	1.00±0.01	0.63±0.01	0.33±0.02	1.59±0.03
1.000	200	0.50±0.02	1.00±0.01	0.55±0.02	0.36±0.02	1.82±0.07
0.950	100	0.48±0.03	1.00±0.01	0.57±0.03	0.29±0.03	1.75±0.09
0.875	100	0.48±0.02	1.00±0.01	0.51±0.03	0.28±0.03	1.96±0.12
0.800	100	0.51±0.03	1.00±0.01	0.55±0.07	0.25±0.08	1.82±0.23
0.750	100	0.45±0.03	1.00±0.01	0.43±0.03	0.16±0.07	2.33±0.16
0.700	100	0.47±0.03	1.00±0.01	0.55±0.07	0.12±0.05	1.82±0.23
0.625	100	0.48±0.04	0.58±0.03	0.63±0.03	0.40±0.03	0.92±0.06
0.500	100	0.51±0.03	0.25±0.03	0.65±0.06	0.61±0.01	0.35±0.10
0.375	100	-	0.16±0.03	0.44±0.05	0.55±0.03	0.36±0.07
0.250	100	-	0.14±0.03	0.25±0.05	0.48±0.04	0.56±0.16
0.125	100	-	0.11±0.03	0.29±0.04	0.48±0.03	0.38±0.12

strong as the nonlocal shadowing effect, and has negligible influence on the wavelength exponent when shadowing is present.

The wavelength selection is a result of the shadowing effect due to the angular distribution of the deposition flux. Figure 8.7 is a schematic diagram showing the concept of "shadowing length" which gives rise to a quasi-periodic mound structure [57], as higher surface features shadow a nearby region of lower surface heights. For normally incident atoms ($\theta = 0°$), the shadowing length is zero, but as the incident angle increases, the shadowing length also increases. An incident flux with an angular distribution therefore gives rise to a distribution of shadowing lengths. The average value of the shadowing length weighted by the angular flux distribution gives rise to the wavelength selection observed in the simulations. As the surface grows rougher in time, tall surface features get even taller and, consequently, the average shadowing length gets larger, along with the wavelength.

The behavior of the exponent $1/z$ in the simulations is significantly different from the wavelength exponent p. From Table 8.1, $1/z$ can lie between 0.12 to 0.61 depending on the sticking coefficient, whereas the wavelength exponent $p \approx 0.5$ whenever there is wavelength selection. The fact that the wavelength exponent is independent of the sticking coefficient (for $s_0 > 0.5$) could suggest that these mounded surfaces may have a "universal" behavior when regarding wavelength selection. However, there is clearly no such universal behavior for the evolution of the lateral correlation length governed by the exponent $1/z$, which depends strongly on the sticking coefficient. Experimentally, the value of $1/z$ reported in the literature scatters between 0.13 to 0.85 [28, 53, 59, 60, 93, 94, 116, 154, 187]. Therefore, it is reasonable to conclude

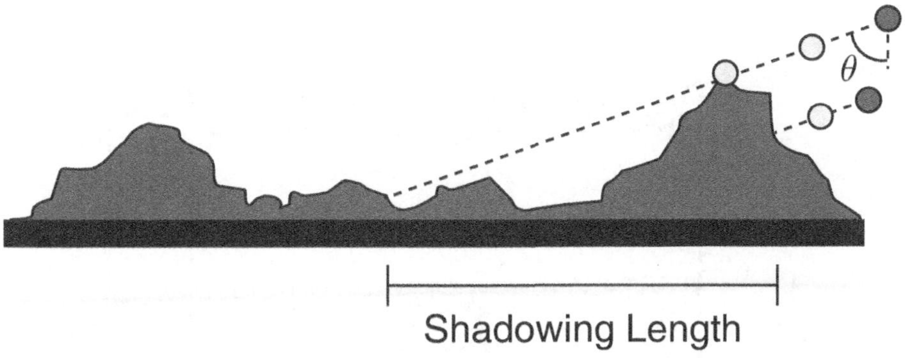

Fig. 8.7. This diagram illustrates the effect of shadowing from obliquely incident atoms and the definition of a "shadowing length" that gives rise to wavelength selection. Atoms strike the surface with an incident oblique angle θ.

that the value of $1/z$ is not universal and strongly depends on deposition conditions.

In addition, the growth exponent β associated with the temporal evolution of the interface width behaves in a manner consistent with shadowing in a solid-on-solid model. Consider a surface grown under the influence of shadowing, and consider a point (x, y) on the surface that is shadowed. By definition, if a surface point is shadowed, it receives little or no incident particle flux, and as a result its growth rate is significantly smaller than the growth rate of the mean surface height. Thus, after sufficient deposition time, the shadowed surface height $h(x, y) \ll \overline{h}$ as a result of the large difference in growth rates. It follows that, because the mean height never stops growing during the deposition, the term in the interface width w involving the shadowed surface height is approximately $\left(h(x, y) - \overline{h}\right)^2 \approx \left(\overline{h}\right)^2$. Eventually, terms in the interface width involving unshadowed surface heights are negligible when compared to $\left(\overline{h}\right)^2$, which gives $w \sim \sqrt{\left(\overline{h}\right)^2} \sim \overline{h}$. The mean height is linear in deposition time, therefore this argument implies that the exponent $\beta = 1$ in strong shadowing growth. Conversely, with strong reemission, shadowed surface heights can grow at a rate similar to the mean height, which may allow for a smaller value for β. The simulation results in Table 8.1 confirm this theoretical prediction. Also, the simulation results predict that reemission begins to become significant when $s_0 \approx 0.7$, at the point where β begins to decrease. Because reemission tends to smooth the surface, strong reemission will slow the growth of the interface width, thereby decreasing the value of β. Reemission becomes the dominant growth effect when $s_0 < 0.5$, where the surface is no longer mounded due to a lack of wavelength selection.

To examine the validity of the MC simulation results, we compare the simulation results to experimental surfaces that have been deposited using sputter deposition and chemical vapor deposition [122, 123]. Both of these

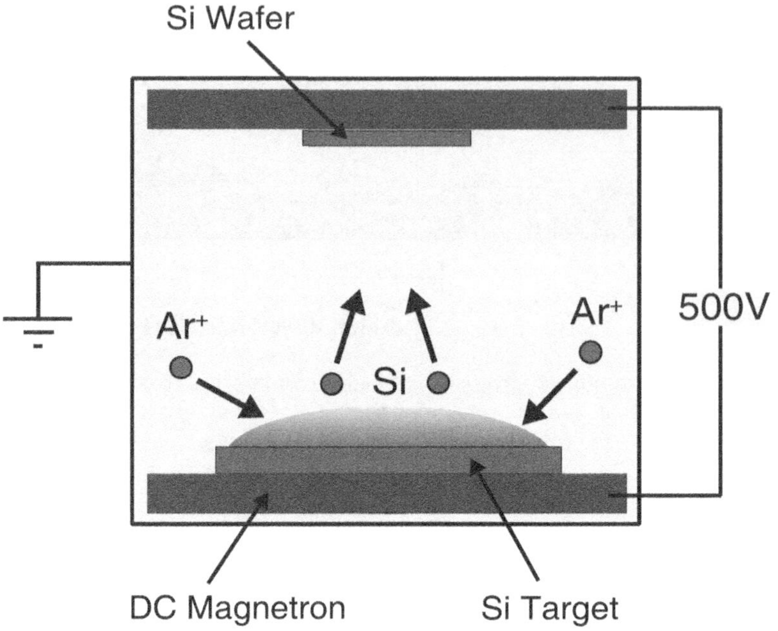

Fig. 8.8. Diagram of the dc magnetron sputtering system used to deposit Si on a Si(100) substrate.

deposition techniques introduce an angular flux on the substrate that is required for shadowing to take place. In addition, in these experiments, silicon was used as a source material because silicon films, under suitable deposition conditions, can be made amorphous. Crystalline effects were ignored in the MC simulations, so amorphous films are more appropriate to compare simulation results with experiment.

A dc magnetron sputtering system was used to deposit amorphous Si on an initially flat Si(100) substrate. A schematic of the deposition system is shown in Fig. 8.8. In all depositions, a power of 200 watts and an Ar pressure of 2.0×10^{-3} torr was used. Depositions ranging from 7.5 to 960 min were performed at a deposition rate of approximately 8 nm/min. The surfaces were imaged using atomic force microscopy (AFM), and images of these surface profiles are given in Fig. 8.9. For each deposition, statistics from four different AFM scans have been averaged, and the results are depicted in Fig. 8.10. The analysis gives $p = 0.51 \pm 0.03$, $1/z = 0.38 \pm 0.03$, $\beta = 0.55 \pm 0.09$, and $\alpha = 0.69 \pm 0.09$. Even though shadowing is present in this deposition, $\beta < 1$ because reemission is also significant. The values of p, $1/z$, β, and α are consistent with the results of the MC simulations with a sticking coefficient $s_0 \approx 0.65$, well within the regime of wavelength selection as predicted by simulation results.

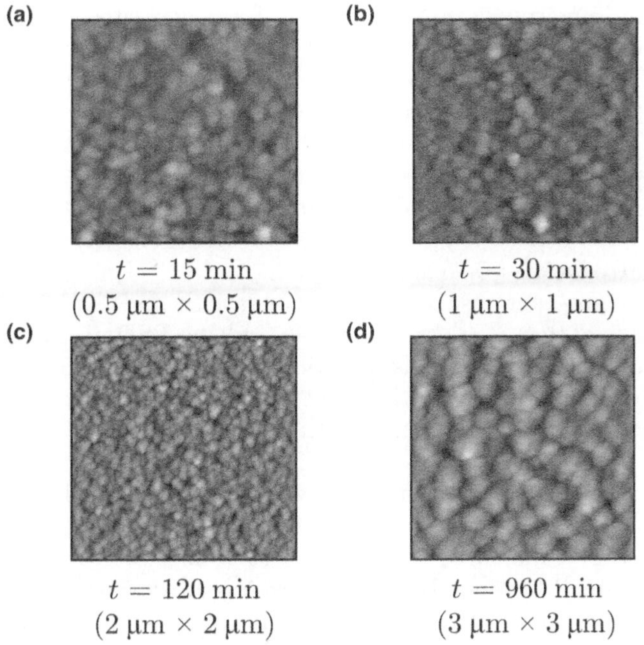

Fig. 8.9. Atomic force microscopy (AFM) images of sputtered Si on Si. Each image represents the surface profile at a different deposition time t. The size of each image is given in parentheses [122].

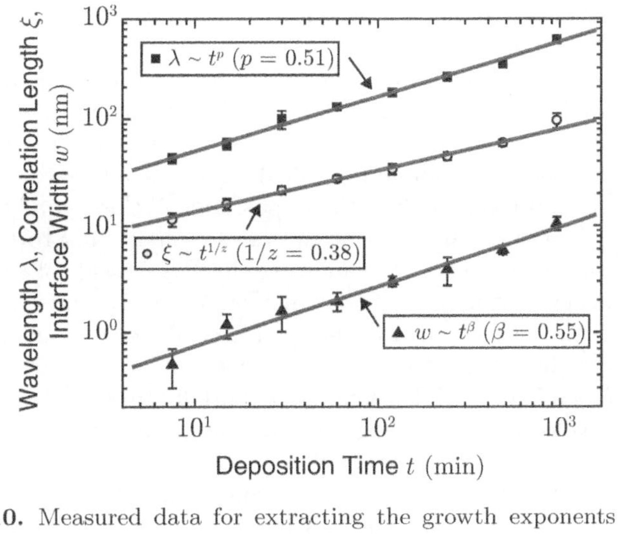

Fig. 8.10. Measured data for extracting the growth exponents for sputtered Si on Si (see Fig. 8.9). The extracted values for the exponents are $p = 0.51 \pm 0.03$, $1/z = 0.38 \pm 0.03$, and $\beta = 0.55 \pm 0.09$ [122].

In addition, amorphous SiN films have been deposited using a plasma enhanced CVD (PECVD) procedure [60]. The front side of Si(100) wafers, which were RCA cleaned prior to deposition, were used as the substrate surface. Depositions were performed at a substrate temperature of 150°C and times ranging from 10 to 180 min at a deposition rate of 5.72 nm/min. The AFM images of the SiN surface profiles are given in Fig. 8.11. The time evolution of the wavelength λ, lateral correlation length ξ, and interface width w are plotted in Fig. 8.12. The analysis gives $p = 0.50 \pm 0.06$, $1/z = 0.28 \pm 0.02$, $\beta = 0.41 \pm 0.01$, and $\alpha = 0.75 \pm 0.04$.

The most important result of these simulations and experimentally deposited surfaces is that $p \neq 1/z$ in general, and the PSD of the surface profiles should not scale in time. This behavior is clearly seen in Fig. 8.13, which contains various PSD curves extracted at different stages in the evolution of surfaces created in a MC simulation with $s_0 = 1$ and $D/F = 100$, and from sputtered Si surfaces described earlier. The PSD curves are scaled so their peaks coincide, which results in the wavenumber axis multiplied by a factor of $\lambda \sim t^p$. Because the peak position defines the value for the wavelength, scaling the peaks of the curves corresponds to scaling the surfaces according to long-range (small wavenumber) behavior. A clear deviation is observed in the spread of the curves. The behavior of the PSD for larger wavenumbers corresponds to the short-range behavior of the surface as represented by the lateral correlation length. Because $p \neq 1/z$ for these surfaces, these length scales do not evolve at the same rate, which leads to the behavior seen in Fig. 8.13. In the scaled curves, from Sect. 4.3, the spread is proportional to $t^{-1/z}t^p = t^{p-1/z}$, and because $p > 1/z$ in these examples, the widths of the scaled curves increase with time. Therefore, the nonlocal effects that lead to mound formation do not allow the system to scale, and the system loses its dynamic scaling behavior.

In the MC simulations and experimental surfaces that exhibit wavelength selection, the wavelength exponent $p \approx 0.5$ when wavelength selection is present, which suggests that the growth process responsible for the value of the wavelength exponent is common to all depositions analyzed. One such growth effect is the noise inherent to the deposition. A closer analysis of the shadowing effect suggests that noise is required for shadowing to take place when the initial surface is flat. The shadowing effect is a result of the competition between surface features of different heights to receive incident particle flux. The noise in the system allows some surface features to randomly grow taller than others, which leads to shadowing. Without noise, starting from a flat substrate, no surface heights would preferentially grow taller than others, eliminating shadowing. This suggests that the nature of the noise in the system has an effect on the value of the wavelength exponent.

A theoretical argument for $p = \frac{1}{2}$ can be constructed using results of the needle model discussed previously. For a 1+1 dimensional surface grown under shadowing, ignoring lateral growth on the surface, Meakin et al. [107] showed that the linear concentration of unshadowed mounds $c(t)$ in such a model

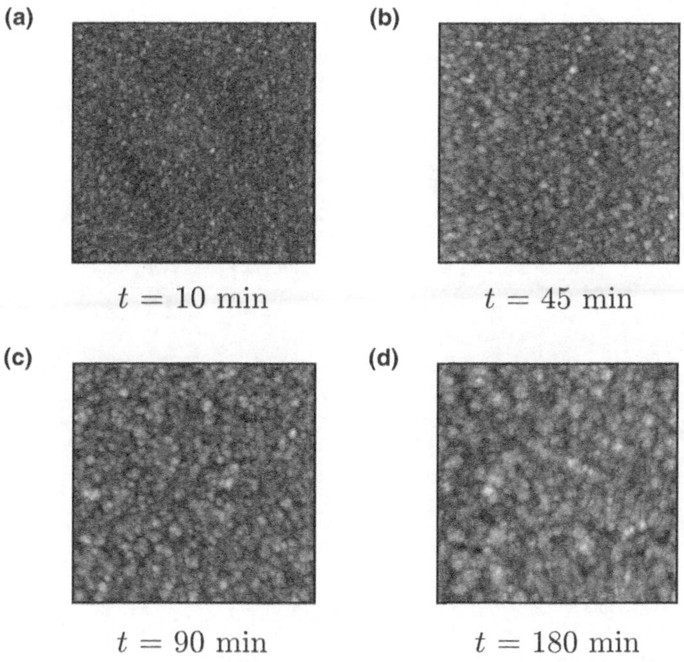

Fig. 8.11. Atomic force microscopy (AFM) images of PECVD SiN. Each image represents the surface profile at a different deposition time t. The size of each image is 2 μm × 2 μm [122].

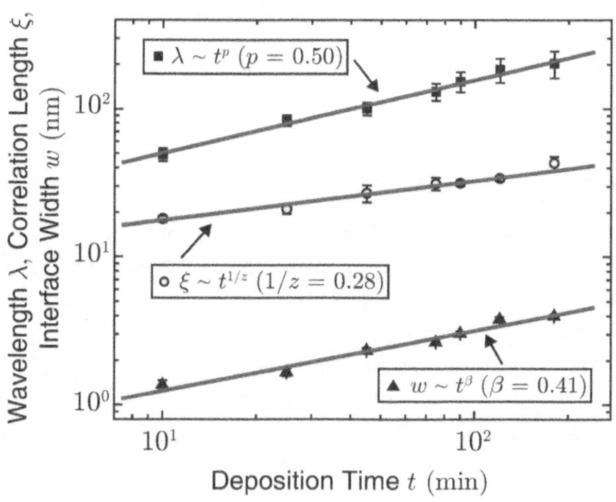

Fig. 8.12. Measured data for extracting the growth exponents for PECVD SiN (see Fig. 8.11). The extracted values for the exponents are $p = 0.50 \pm 0.06$, $1/z = 0.28 \pm 0.02$, and $\beta = 0.41 \pm 0.01$ [122].

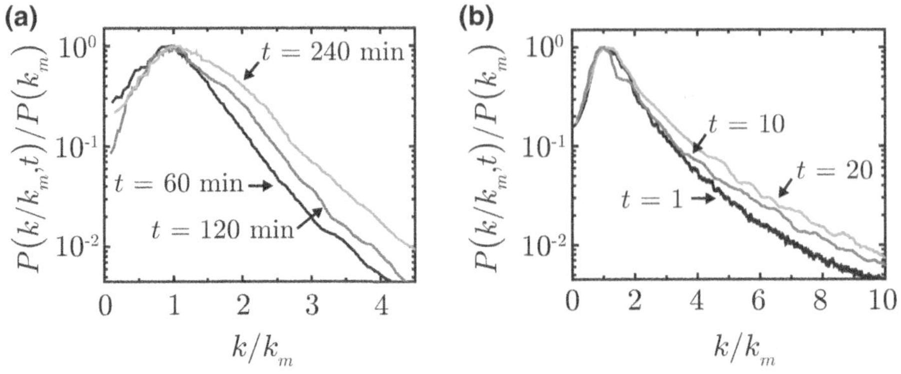

Fig. 8.13. Scaled PSD curves of (**a**) sputter deposited (experimental) surfaces and (**b**) simulated ($s_0 = 1$, $D/F = 100$) surfaces scaled according to peak position. The spread of the curves does not scale, consistent with the prediction of the breakdown of dynamic scaling [122].

behaves as

$$c(t) \sim t^{-1/2}. \tag{8.1}$$

This result is derived from the condition that unshadowed mounds grow according to a Poisson process, which implies that individual mound heights perform a random walk about their mean. This is a reasonable assumption because unshadowed mounds experience the full incident particle flux, which is subject to a Gaussian noise distribution.

Using a simple geometric argument, $c(t)$ can be related to the wavelength λ. For a 1+1 dimensional surface, if the surface is of linear size L, and the average distance between mounds is λ, then there are L/λ mounds on the surface. Similarly, by the definition of $c(t)$, there are $c(t)L$ mounds on the surface. This implies that $c(t)L \sim L/\lambda$, or

$$c(t) \sim \lambda^{-1}, \tag{8.2}$$

from which $p = \frac{1}{2}$ follows. A similar argument holds in 2+1 dimensions, recalling that $c(t)$ is a linear density of mounds. The total number of mounds on the surface can be represented as $(c(t)L)^2$ and $(L/\lambda)^2$ in 2+1 dimensions, which again leads to $p = \frac{1}{2}$.

Even though this argument correctly predicts that $p = \frac{1}{2}$, it is based on a model that ignores the lateral growth of mounds, or, similarly, ignoring the lateral correlation length governed by the exponent $1/z$. However, from simulation results, the behavior of p and the behavior of $1/z$ do not appear to be correlated under different deposition conditions. It therefore seems reasonable that p and $1/z$ are independent in this context, with p determined by the noise and $1/z$ determined by deposition conditions such as the sticking coefficient and strength of diffusion. This argument is far from a proof for the general behavior of the wavelength exponent p, and further work is needed to fully quantify the behavior of the wavelength exponent under various deposition conditions.

In addition, from Sect. 3.5, when a surface obeys dynamic scaling, the growth exponents are related in a specific way, namely,

$$z = \frac{\alpha}{\beta},$$

or, equivalently,

$$\frac{1}{z} = \frac{\beta}{\alpha},$$

which is a more convenient form because the exponent $1/z$ is measured directly from the lateral correlation length. This relation should no longer hold for surfaces grown under the influence of shadowing because dynamic scaling no longer holds for these surfaces. For the experimental sputter deposition,

$$\frac{1}{z} = 0.38 \pm 0.03, \quad \frac{\beta}{\alpha} = 0.80 \pm 0.17,$$

which do not agree with the relationship $1/z = \beta/\alpha$ within experimental error. Also, for the experimental CVD,

$$\frac{1}{z} = 0.28 \pm 0.02, \quad \frac{\beta}{\alpha} = 0.49 \pm 0.03,$$

which also do not agree with the relationship $1/z = \beta/\alpha$ within experimental error. For the MC simulation results in Table 8.1, the last two columns of the table give values for $1/z$ and β/α, respectively, for comparison. Note that when wavelength selection is dominant (for $s_0 \geq 0.5$), there is a significant difference between $1/z$ and β/α. However, when reemission is strong enough to cancel wavelength selection, $1/z \approx \beta/\alpha$ within measurement error. These results suggest that when the shadowing effect is sufficiently suppressed during deposition, the surface becomes self-affine and obeys the dynamic scaling hypothesis. Note, however, that simply investigating the validity of the relation $1/z = \beta/\alpha$ is not sufficient to claim a breakdown of dynamic scaling alone. It is simply an observation that logically follows when dynamic scaling has been broken, as a result of $p \neq 1/z$ under shadowing.

A number of previous studies on the effects of shadowing [155, 177] and reemission [184] did not examine quantitatively the behavior of the time evolution of the wavelength λ. It is important to note that some authors have used the variable p to describe the time evolution of the lateral correlation length as opposed to wavelength selection. Using a model based on the Huygens principle (HP), Tang et al. [155] examined the evolution of the lateral correlation length ξ of simulated surfaces. The exponent $1/z$ associated with the lateral correlation length depends on the initial surface configurations, and ranges from $\frac{1}{4}$ to 1. However, under the HP, mounds grow next to each other without gaps [7, 155], and the spacing between mounds is the same as the mound size, or $\xi = \lambda$, which implies $p = 1/z$. A continuum model presented in Yao and Guo [177] accounted for shadowing during the growth process which predicted $1/z = 0.33$, consistent with simulation results under the specific condition of $s_0 = 1$.

Also, it is noted that previous work on the dynamic scaling behavior of surfaces grown by MBE under a step diffusion barrier utilized similar analysis techniques to those used in this work. In particular, Siegert [140] showed that, under certain conditions in MBE growth, the surface can be quantified by two length scales that do not evolve at the same rate, similar to the discussion of the wavelength λ and correlation length ξ presented here. Also, Moldovan and Golubovic [113] showed that for simulated surfaces grown under MBE, the height–height correlation function does not exhibit time-dependent scaling. The PSD can be related to the Fourier transform of the height–height correlation function, therefore analyzing the time-dependent scaling behavior of the height–height correlation function is similar to analyzing the time-dependent scaling behavior for the PSD. However, both these papers focused solely on MBE, which is governed by a local step-barrier diffusion effect which can be modeled by a local continuum equation. The shadowing and reemission effects are nonlocal, and lead to a markedly different surface morphology than is created in MBE.

8.2.2 Competition Between Shadowing and Reemission

Although shadowing and reemission are both nonlocal effects, they are opposites in the sense that shadowing enhances roughness and reemission reduces roughness. The effectiveness of the reemission effect depends very much on the value of the sticking coefficient s_0, which may vary from 0 to 1 [184]. It is therefore interesting to study quantitatively in more detail the competition between shadowing and reemission as we vary the sticking coefficient [92]. Figure 8.14 illustrates the results of this competition. Under shadowing, the surface tends to roughen quickly if the sticking coefficient is large (weak reemission), and roughen more slowly if the sticking coefficient is small (strong reemission).

As depicted in Fig. 8.15, we assume that when an atom first strikes the surface (inset point A), it has a sticking coefficient of s_0. When an atom bounces and strikes the surface at another position (inset point B), in the data presented, the sticking coefficient is taken to be unity. This is called a first-order reemission model [34]. In sputter deposition, for example, the incident atom energy may be high when it first strikes the surface, and the sticking coefficient may significantly differ from 1. However, upon collision with the surface, the atom loses kinetic energy and the reemitted atom has a higher probability (s_1) to stick the second time.

For a more quantitative discussion of the competition between shadowing and reemission, we now consider the case of a deposition where the incident

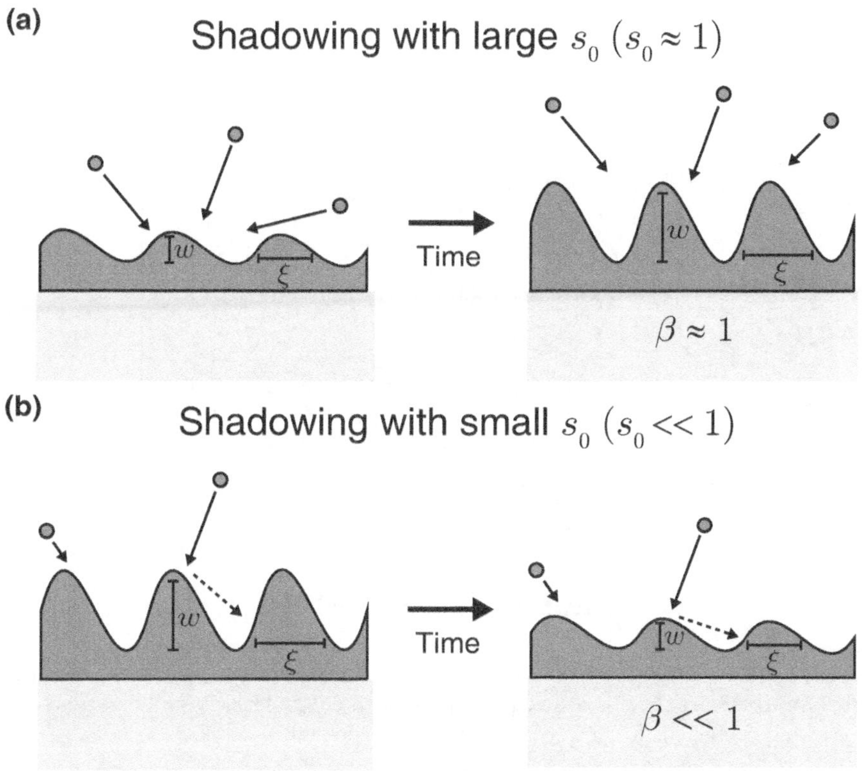

Fig. 8.14. Illustration of the competition between shadowing and reemission effects. For large sticking coefficients, the surface roughness is enhanced by shadowing. For small sticking coefficients, reemission is strong enough to counter shadowing and make the surface roughen more slowly. The behavior in part (**b**) would not be observed in growth from a flat substrate as the roughness does not usually decrease with time in such a deposition. However, beginning from a very rough surface, reemission can smooth the surface if s_0 is small.

flux has a distribution of $\cot\theta$ [92]. Results of solid-on-solid MC simulations for different values of the sticking coefficient s_0 and different diffusion strengths D/F are pictured in Fig. 8.15. We see that varying the sticking coefficient qualitatively changes the behavior of the growth exponent β, whereas changing the strength of surface diffusion results only in a small quantitative change in β. For small sticking coefficients, surface diffusion has little effect on the value of β, and the surface tends to roughen slowly. As was shown with similar simulation results in Sect. 8.2.1, in this regime, the simulated morphology is self-affine, and should obey dynamic scaling during growth. No mounds are formed for small s_0, and wavelength selection does not exist. In the high sticking coefficient regime, surface diffusion has little effect on the value of β and the surface tends to roughen linearly with time, which implies $\beta \approx 1$.

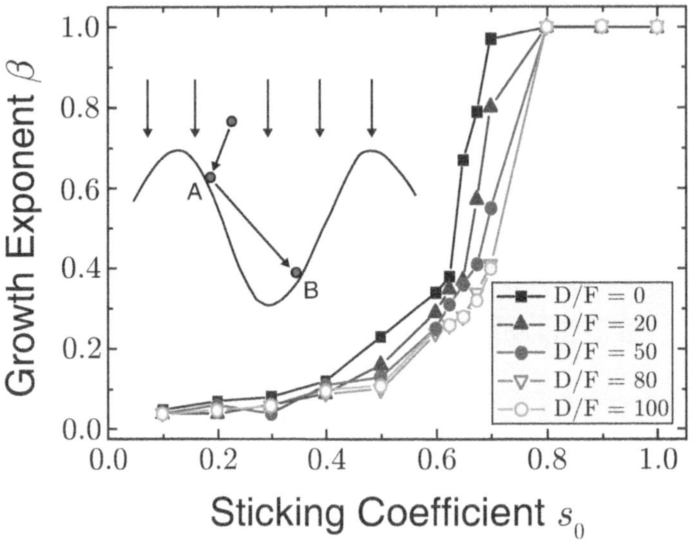

Fig. 8.15. Monte Carlo simulation results for the growth exponent β as a function of sticking coefficient s_o for a flux proportional to $\cot\theta$. These results assume that the sticking coefficient for the second strike, s_1, is equal to 1 [92].

As shadowing is dominant in this regime, mounds are formed on the surface, and the surface is no longer self-affine. For intermediate values of the sticking coefficient, changes in β can be observed as a result of surface diffusion, but shadowing and reemission have a much larger effect on the growth exponent β than does surface diffusion. The results shown in Fig. 8.15 with a $\cot\theta$ incident flux distribution behave similarly to the results from a $\cos\theta$ distribution as given in Table 8.1.

At this point it is interesting to note that for an etching process, the relationship between β and s_0 is opposite to the behavior observed for a deposition process. Etching differs from deposition in the sense that incident particles remove atoms from the substrate in an etching process, whereas incident particles accumulate at the substrate interface in a deposition process. Therefore, if a surface is mounded in an etching process, the mounds will be preferentially etched away because they receive a greater flux of etchants. This leads to a much more complicated behavior for mounds on the surface under etching, which has not been extensively studied previously. Certain investigations [32, 185] suggest that the average hole separation λ, which would be the analogue of mound formation under etching, behaves as $\lambda \sim t^{\gamma}$, where $\gamma \approx 1$.

For a large sticking coefficient, etching leads to a smooth growth front. Etchants arrive and react with the surface atoms to form volatile species.

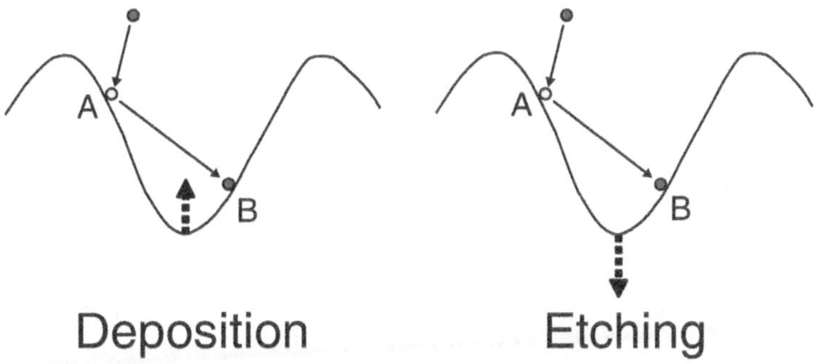

Fig. 8.16. Effect of a small sticking coefficient on deposition and etching processes under shadowing. Reemission tends to smooth the surface during deposition, but tends to roughen the surface during etching.

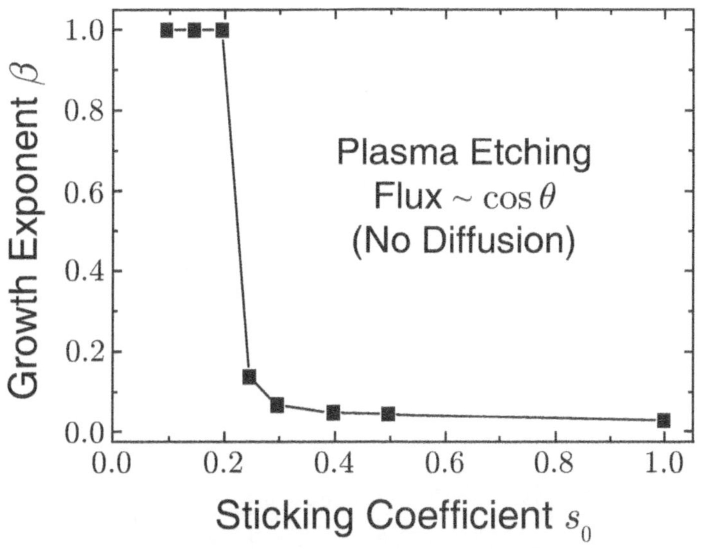

Fig. 8.17. Simulated growth exponent β as a function of the sticking coefficient s_0 for the plasma etching of a surface, where incident particles are modeled to follow a $\cos\theta$ angular distribution. Diffusion is not considered in these simulations, and as was the case in Fig. 8.15, diffusion does not change the qualitative nature of the relationship between β and s_0 [92].

The volatile species then leave the surface and leave behind vacancies on the surface. For plasma etching, the incident flux can be modeled by a $\cos\theta$ distribution. If the sticking coefficient s_0 is unity then, as time evolves, shadowing would tend to smooth the surface to give $\beta \approx 0$, just the opposite of the behavior for a deposition. During etching, tops of mounds are removed by the etching flux, and the valleys on the surface remain because they are not exposed to as much etching flux.

However, if the sticking coefficients are small, as in many etching processes using etchants such as CF_4, the etchant can reach the valleys of the surface through reemission. Therefore, atoms in valleys can be removed, leading to deeper valleys and a rougher surface, which results in a larger value of β [17, 126, 185]. This behavior is pictured in Fig. 8.16. Often, for small sticking coefficients, the measured value of β is approximately 1. Simulated results for the growth exponent β as a function of the sticking coefficient s_0 are shown in Fig. 8.17 in the absence of surface diffusion. As was the case for deposition, we expect that surface diffusion does not change the curve qualitatively. The graph indicates that the surface becomes very rough when the sticking coefficient s_0 is smaller than 0.2, and becomes much smoother when the sticking coefficient s_0 is larger than 0.2.

9

Ballistic Aggregation Models

Ballistic aggregation models, as introduced in Sect. 7.2.2, allow for overhangs on the surface as a result of allowing particles to attach themselves to the aggregate if a neighboring lattice point is occupied. This simple modification of a solid-on-solid model can lead to an interesting behavior of the resulting aggregate. We consider only on-lattice ballistic aggregation in this chapter, where the particles can aggregate in the constraints of a cubic lattice. Other techniques allow for aggregation in any configuration, which is called off-lattice ballistic aggregation.

9.1 Comparison to Solid-on-Solid Models

When considering the modeling of films from a flat initial substrate, one can utilize both solid-on-solid models and ballistic aggregation models to predict surface roughening behavior. Depending on the experimental conditions to be modeled, it is the choice of the investigator which model to utilize appropriately. Ballistic aggregation models have been used to model film growth from a flat substrate [9, 105, 111, 176, 181], and the models presented in the previous chapter are examples of solid-on-solid models of growth from a flat substrate. The differences in these two types of models for the specific case of growth from a flat substrate are worth discussing, as the models can provide significantly different results.

As has been discussed throughout the text, films grown from a flat substrate are quantified with surface statistics such as the interface width w, mean height $\overline{h}$, and lateral correlation length ξ. These statistics are defined assuming that the surface itself can be described by a single-valued function of position $h(\mathbf{x})$. For a solid-on-solid model, because of the nature of the "drop down" aggregation, the surface is well-defined in terms of a single-valued function $h(\mathbf{x})$. However, under a ballistic model which allows for overhangs, the surface cannot be fully quantified with a single-valued function for the height, because such a function would neglect overhangs and the underlying structure

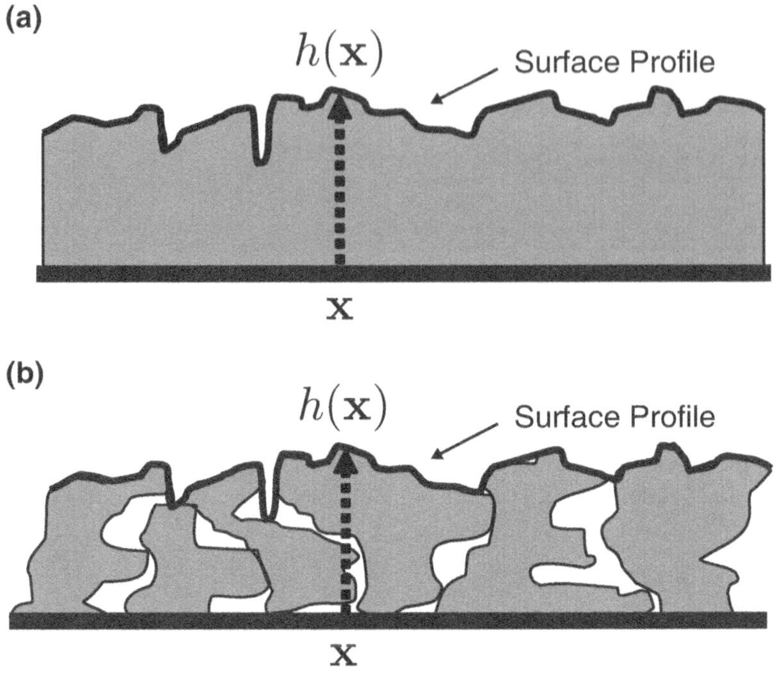

Fig. 9.1. Identical surface height profiles in (**a**) a solid-on-solid model, and (**b**) a ballistic aggregation model. No overhangs are allowed in the solid-on-solid growth model, and the height $h(\mathbf{x})$ is a single-valued function. Overhangs are allowed in the ballistic growth model, and the height $h(\mathbf{x})$ is defined as the highest surface point at the position $\mathbf{x}$.

of the film. Therefore, statistics such as the interface width that are defined in terms of a single-valued height function are ill-defined for a surface formed under ballistic aggregation, and suitable consideration must be paid to this discrepancy. One cannot simply use the traditional surface statistics to describe these surfaces without an adequate explanation of how the statistics were obtained.

The most straightforward way to extract a description of the surface is to simply take $h(\mathbf{x})$ to be the maximum surface height at any position on the surface, which would be similar to observing the surface profile from above. This method is pictured in Fig. 9.1. Experimentally, measurement techniques such as atomic force microscopy (AFM) and scanning electron microscopy (SEM) take a "top view" image of the surface that can be regarded as single-valued. However, when comparing these results with the results of a ballistic aggregation model, it is possible that surface statistics will depend on the size of the tip used experimentally, and a similar consideration must be given to simulated surface statistics. If the experimental tip size is large compared

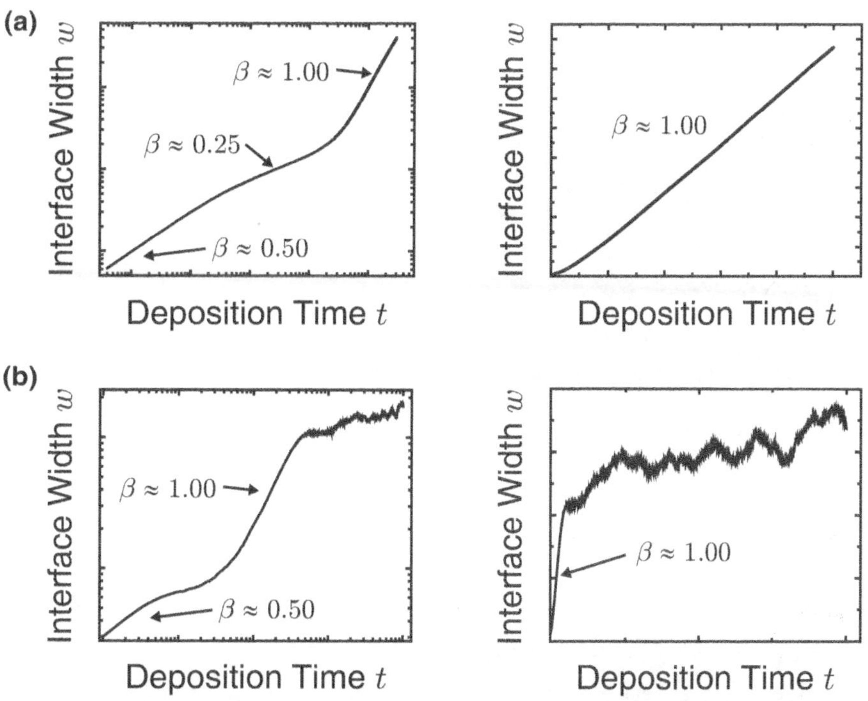

Fig. 9.2. Comparison of the evolution of the interface width w plotted in both linear and log–log scales under identical conditions for (**a**) a solid-on-solid model and (**b**) a side-sticking ballistic aggregation model. Both simulations are under a $\cos\theta$ incident flux distribution with surface diffusion strength $D/F = 100$. Initially, $\beta \approx 0.5$ due to random growth, with a short regime of $\beta \approx 0.25$ where diffusion is the dominant growth mechanism. After a significant number of particles aggregate, $\beta \approx 1$ when geometrical shadowing dominates. The roughness behavior agrees until large times, when presence of overhangs leads to a noisy roughness behavior in the ballistic aggregation model that is absent from the solid-on-solid model.

to an average feature size, experimental data will not reflect the nature of overhangs that may be present in a simulated profile.

These effects are most pronounced in a measurement of the surface roughness. Figure 9.2 depicts the roughness evolution of an intially flat surface under the influence of a $\cos\theta$ flux distribution with diffusion strength $D/F = 100$ and sticking coefficient $s_0 = 1$ for (a) a solid-on-solid model, and (b) a ballistic aggregation model. The roughness behavior is similar except at long simulation times, where the roughness from the ballistic model does not necessarily saturate, but grows at a much smaller rate than previously and is much noisier. This occurs because overhangs somewhat randomly cover up the low-lying areas of the film, significantly changing the measured mean height from above and causing the roughness to fluctuate. This fluctuation is clearly not due

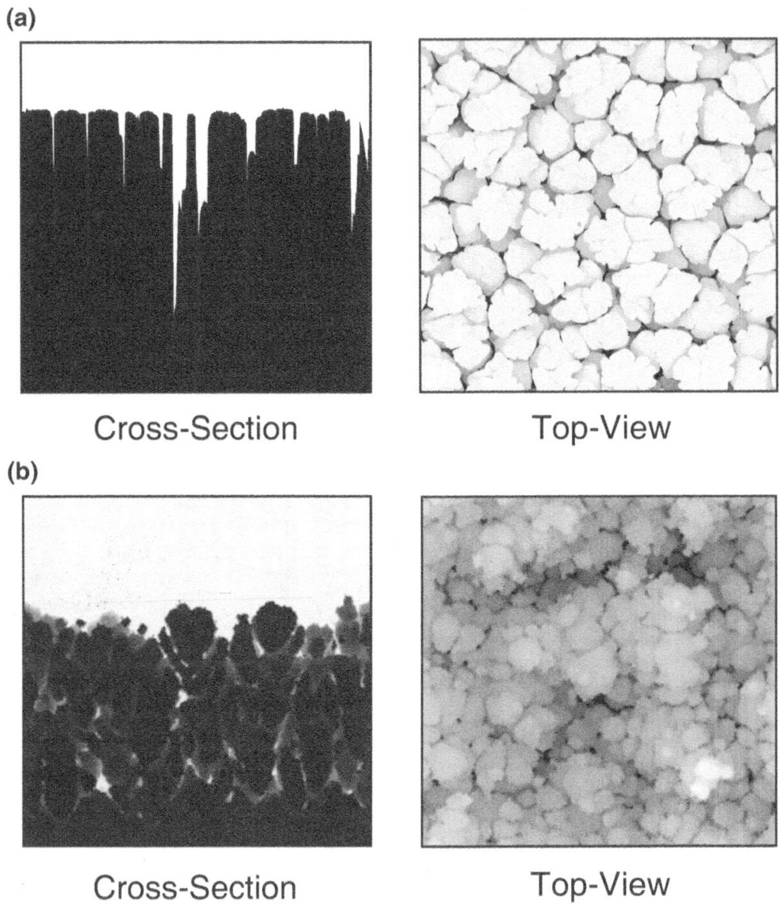

Fig. 9.3. Cross-section and top-view images of the surfaces with roughness curves pictured in Fig. 9.2. In (**a**), the solid-on-solid model has no overhangs, and leads to a sharp columnar structure. In (**b**), the ballistic aggregation model allows for overhangs, and the surface is more porous.

to a natural error in the roughness from the finite size of the lattice, as the roughness grows much more smoothly at shorter times when overhangs are not significant. In the solid-on-solid model, low-lying areas of the film that are shadowed never receive particle flux and are always visible from above, which leads to a smooth roughness evolution. The low-lying areas are still present in the ballistic model, as they are responsible for the rapid roughness growth at intermediate times due to geometrical shadowing, but are not measured at long times because of the single-valued definition of the height required by the roughness. The cross-section and top-view images of these surfaces in included in Fig. 9.3.

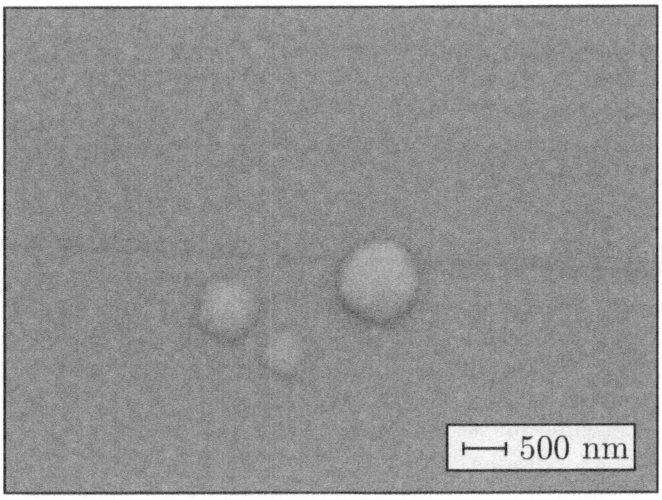

Fig. 9.4. Thermally deposited Si film exhibiting nodular defect formation. The thickness of the film is approximately 2.5 μm. Image courtesy D.-X. Ye.

9.2 Intrinsic Nodular Defects

One example of the use of a ballistic aggregation model is in the growth of nodular defects in films. Nodular defects are mushroom-shaped growths that have been observed experimentally in the attempted growth of uniform thin films by vapor deposition [44, 79, 134, 151]. One would expect to obtain a uniform film in processes that utilize normally incident flux, such as thermal evaporation or electron beam deposition. However, in the growth of relatively thick films, experiments have observed these nodular defects forming during the deposition. If an initial substrate is used that is not sufficiently cleaned, dust or other particles can act as seeds and produce similar mushroom-shaped extrinsic nodular defects. However, experimental results indicate that these defects sometimes do not begin to grow at the substrate interface, but rather naturally occur as a consequence of the deposition process. Figure 9.4 shows an amorphous Si film deposited by normal incidence thermal evaporation that exhibits this intrinsic nodular defect formation. In applications where the uniformity of the film is important, such as optical coatings, formation of these defects is undesirable. Limited modeling has been performed on the formation of these defects, but previous work [37, 79] suggests that ballistic aggregation can provide some insight into nodular defect growth.

To investigate the intrinsic nodular defect growth inherent to the deposition process, we can investigate surfaces formed under the ballistic aggregation of a normally incident particle flux on an initially flat substrate. The specific simulations presented are carried out on three-dimensional lattices of dimension $256 \times 256 \times 4096$ and $512 \times 512 \times 4096$ to investigate both nodular defect growth behavior and the behavior of surface roughness with deposition time. Because the nature of the particle flux is uniformly normal, the most prominent variable in the modeling of this growth is the strength of surface diffusion, which will tend to make the surface smoother and counter the growth of nodular defects.

In addition to surface diffusion, self-shadowing is an important growth effect in the presence of nodular defects, which refers to the local shadowing of the film beneath and surrounding the defect. There is no geometrical shadowing because the flux has no angular distribution, however, because this is a ballistic model, overhangs are allowed, which can shadow the film directly beneath them. As a result, one should expect a complicated behavior of the surface roughness as these nodules begin to form under the influence of self-shadowing. Nodular defect growth will help quantify the behavior of β under self-shadowing conditions.

Figure 9.5 includes the behavior of the interface width for various strengths of surface diffusion. The behavior of the interface width indicates that the growth can be separated into three regimes: the initial regime of random growth that arises because particles will aggregate at random positions on a flat surface under a normal flux, the formation of nodular defects at random sites due to the noise inherent to the deposition, and the saturation of the surface with defects where the initial homogeneous film is no longer visible. This behavior is also depicted graphically in Fig. 9.6.

Each region of the growth is characterized by a particular value of β, where the value of β can be measured from the slope of the interface width curve plotted on a log–log scale. As was shown in the context of continuous models, the value of β for random growth is $\beta = 0.50$, and this value is observed in the random region for all values of surface diffusion. This regime dominates when particles are far apart on average, and the effects of ballistic deposition cannot be realized. As particles begin to aggregate atop one another, ballistic effects become important, and nodular defects begin to grow as a result of relatively large events from the random noise.

The nodular growth region is characterized by an increased value of β due to the tendency of nodules to accumulate particles that would have otherwise been incorporated into the film. The growth of nodules in the model is shown in Fig. 9.7a. In this region, the roughness evolution is sensitive to diffusion. Larger values of diffusion tend to inhibit the growth of nodules because diffusion tends to smooth out the film, making it more difficult for events to occur that lead to nodular defect growth. This is also seen in Fig. 9.5 in the sense that it requires more deposited particles, and consequently more time, for defects to form, the larger the strength of surface diffusion. Increasing the strength of surface diffusion reduces the value of β in this region. For small values of surface diffusion, from Fig. 9.5, $\beta \approx 1$, and for surface diffusion strengths $D/F \geq 200$, β is reduced to $\beta \approx 0.65$.

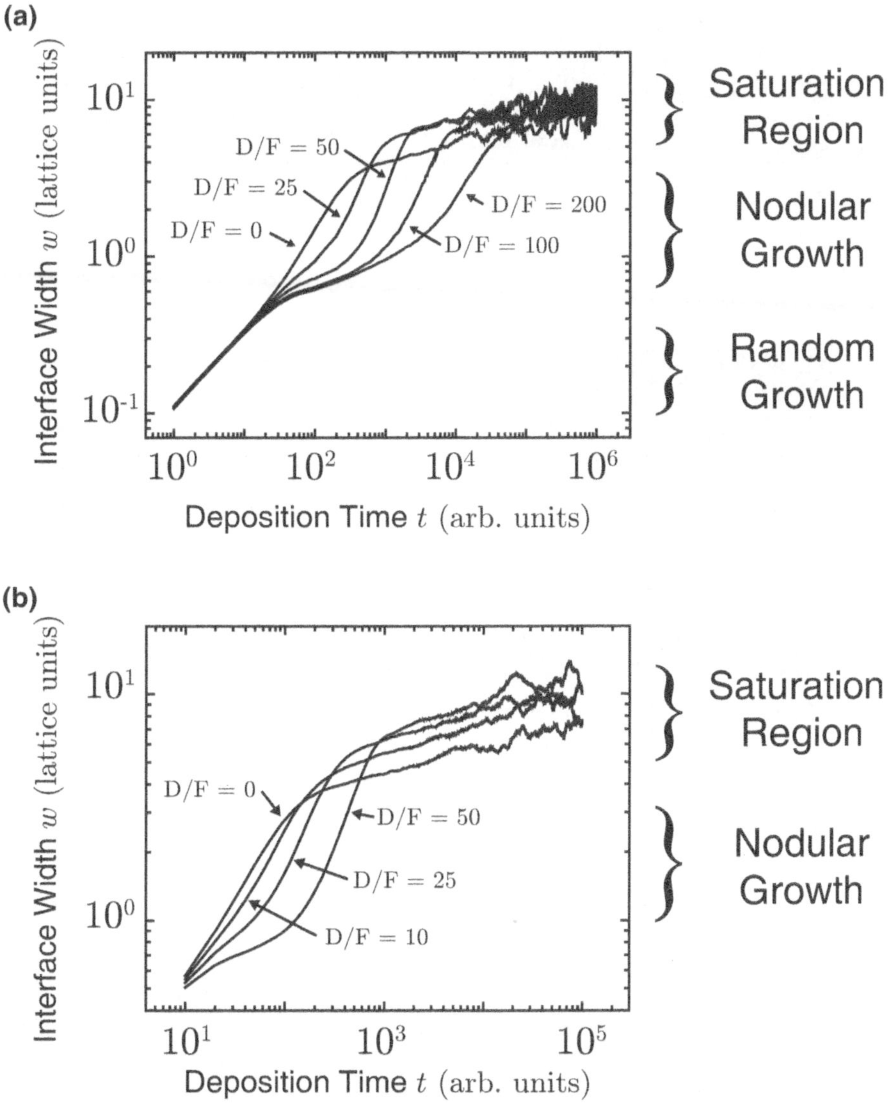

Fig. 9.5. Interface width w versus deposition time t for nodular defect growth by normal incidence flux on an (**a**) $256 \times 256 \times 4096$ lattice and (**b**) $512 \times 512 \times 4096$ lattice. In the nodular growth region, diffusion causes the curves to shift to the right.

This behavior is characteristic of self-shadowing. As previously noted, under strong nonlocal geometrical shadowing due to an angular particle flux, $\beta = 1$. If shadowing is important in nodular defect growth, one should expect to observe similar values for β, as is seen for small values for diffusion in Fig. 9.5. However, self-shadowing can be weakened by surface diffusion,

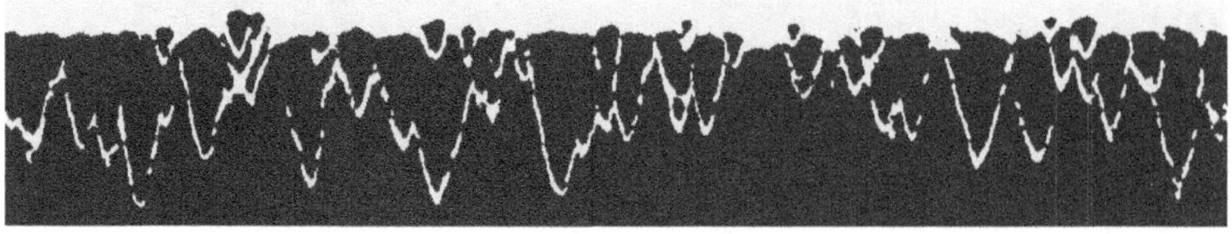

Fig. 9.6. Cross-sectional image of nodular defect growth on a flat film under ballistic deposition at normal incidence, where deposited particles are represented as black. The growth of defects above the initial substrate is realized.

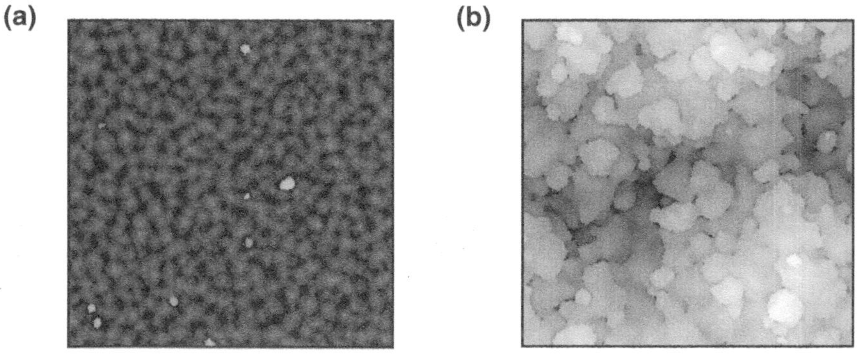

Fig. 9.7. Representative simulation results in (**a**) the nodular growth region with nodular defects and (**b**) the saturation region where the morphology consists of nodules growing atop nodules.

as evidenced by the reduced value of β under strong surface diffusion. This behavior is notably different from conventional shadowing, as $\beta = 1$ for nonlocal shadowing regardless of the strength of diffusion [122]. This suggests that self-shadowing is a weaker effect than nonlocal geometrical shadowing because it can be affected by other local effects including surface diffusion. Even so, self-shadowing still exhibits the characteristic behavior of $\beta = 1$ when surface diffusion is not strong.

In the saturation region, the surface has been saturated with nodular defects. The film has now become a canopy of nodular defects as shown in Fig. 9.7b, and the roughness behavior becomes characteristic of a rough film in a ballistic aggregation model. From Fig. 9.5, the value of β is approximately 0.15, and is constant over different diffusion strengths, and holds even when there is no diffusion, $D/F = 0$. This value of β is then characteristic of this region, where one observes nodules beginning to grow atop nodules as in Fig. 9.6. Experimentally, this region is often not observed because the growth of nodules is relatively slow. For example, the film depicted in Fig. 9.4 is in the nodular growth regime, and is on the order of microns thick, which is the order of magnitude usually desired in the thickness of such films.

The formation of nodules in a ballistic aggregation scheme suggests that nodular growth can be attributed to the growth process itself. To reduce the formation of defects, these results suggest that ballistic sticking, as depicted in Fig. 7.3, must be reduced as this type of sticking promotes the formation of nodules. This can be achieved by using particles in the deposition with larger energies, making each particle less likely to stick on first impact, and more likely to settle on the film through reflections or "drop down" aggregation. In addition, enhancing surface diffusion tends to postpone the formation of defects in the ballistic aggregation model from Fig. 9.5. Overall, this investigation of nodular defect growth in ballistic aggregation exemplifies the utility of MC models, as a relatively simple aggregation scheme can lead to complex behavior without modeling specifics of the depositions.

9.3 Aggregates on Seeds

Extrinsic nodular defects occur when they are induced by the existence of initial extrusions or seeds on the surface [124]. The basic mechanism of extrinsic nodular defect formation can be understood by considering the ballistic aggregation of particles onto a point seed [10, 13, 72, 81, 83, 128, 132]. In this model, normally incident particles randomly rain down towards an initial seed where they are allowed to aggregate. Incident particles follow a straight-line path towards this aggregate, and may attach themselves to the aggregate if their path collides with, or comes sufficiently close to, a deposited particle on the aggregate. This aggregation results in the formation of a fanlike aggregate that exhibits interesting scaling properties. In particular, it has been shown that these aggregates show a self-similar scaling behavior, while possessing a trivial fractal dimension [13, 72]. Most theoretical work and simulation results based on ballistic aggregation onto a seed have been carried out in 1+1 dimensions [10, 13, 81, 83, 128, 132], although certain investigations in 2+1 dimensions have also been reported [72].

Much like the formation of intrinsic nodular defects discussed in the previous section, this type of ballistic aggregation phenomenon is relevant to the vacuum deposition of atoms. Under the high vacuum conditions of many thin film deposition techniques, such as thermal evaporation, the mean free path of particles incident on the substrate is often much larger than the source–substrate separation, allowing for the assumption that the particle trajectories are linear far from the substrate [96]. This type of growth has been observed in the formation of fanlike nodular structures on surface defects which intrude above a nominally flat surface [37, 44, 79, 134, 151], or on prefabricated seeds [179].

In certain vacuum deposition techniques such as oblique angle deposition, it is possible to rotate the incident particle flux during deposition to form "nodular structures" in the form of nanorods from initially nucleated islands, as depicted later in Fig. 9.15a [57]. Oblique angle deposition has been shown to create novel nanorod structures that have unusual optical, magnetic, and electrical properties [29, 30, 46, 48, 49, 50, 52, 56, 65, 66, 67, 69, 75, 84, 85, 136, 139, 157, 158, 164, 173, 180, 188, 189]. The morphology of nanorods is extremely important in these structures, as the control of this morphology leads to better control of the physical properties for potential applications. Both theoretical and experimental investigations into nanorod morphology have centered on global growth effects caused by the nonlocal shadowing of nanorods by adjacent nanorods [29, 56, 57, 58, 164, 189]. However, the growth effects that dictate the shape of individual nanorods is not well understood. In this section, we show that a variation of the incident flux direction during deposition can significantly change the nodular fan structure observed on prefabricated seeds or islands generated during the initial stages of deposition on a flat surface. In particular, the growth of the width of the fan changes dramatically due to a self-shadowing effect as one varies the incident flux [124]. This self-shadowing effect is found to be the dominant factor that controls the growth of the fan width, which can then be used to model nanorod growth during the initial stages of oblique angle deposition.

The simulations discussed in this section are on-lattice ballistic aggregation simulations, performed on a three-dimensional cubic lattice of size $1024 \times 1024 \times 1024$. The number of particles N used in each simulation is $N = 1.5 \times 10^8$. Initially, a seed is defined along an edge of the cubic lattice, and particles are chosen to enter the lattice at a random position, with a trajectory suitably defined to model a uniform flux that makes an angle θ with a reference direction, which we call the normal, as depicted in Fig. 9.8. Aggregate rotation is modeled by rotating the incident particle flux with a change in the azimuthal angle of $\Delta\phi = 0.0288°$ in the trajectory of each particle. The simulation tracks the trajectory of a particle until it becomes part of the aggregate by occupying a lattice point adjacent to a previously occupied lattice point, or until the particle travels past the aggregate, at which point a new trajectory is chosen.

9.3.1 Aggregates Without Diffusion

We begin with the analysis of the ballistic aggregation of particles on a point seed under the conditions of an oblique particle flux and aggregate rotation. Cross-section images of aggregates grown under rotation and without surface diffusion are included in Fig. 9.9. The initial seed is centered at the bottom of each image. For small oblique flux angles, the aggregates tend to form a conical structure due to the inclusion of nearest-neighbor sticking in ballistic aggregation, whereas larger oblique flux angles lead to significant growth on the sides of the aggregate and a deviation from conical aggregation.

To describe the aggregate more quantitatively, we define a polar coordinate system (r, φ) with the origin at the initial seed, where the axis $\varphi = 0°$ corresponds to the vertical growth direction, or the z-axis in a Cartesian coordinate

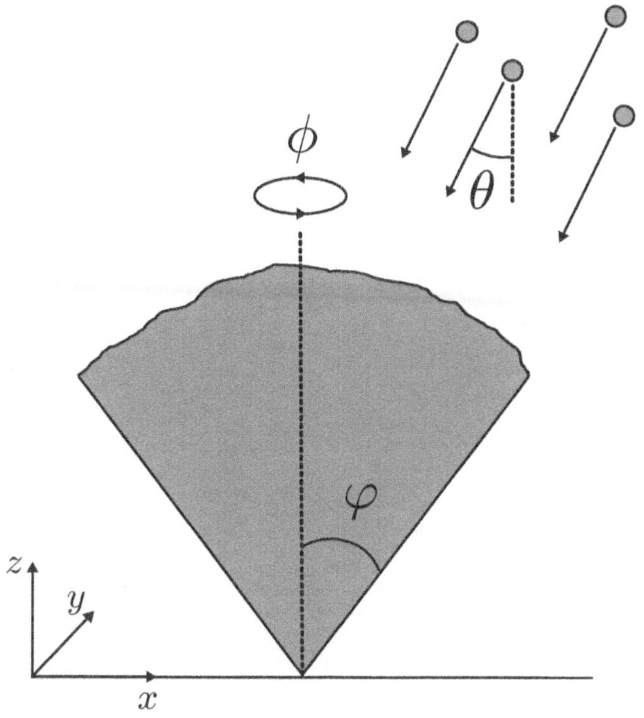

Fig. 9.8. Illustration of the growth of an aggregate on a point seed. The incident flux makes an angle θ with the suitably defined normal, and rotation occurs along the ϕ direction [124].

system. Due to the rotation of incoming particle flux during the simulations, on average, the aggregates are symmetric about the vertical axis, which leads to the use of the radial coordinate r. We focus on the extent to which the aggregates "fanout" from the initial seed, quantified by the maximum angle φ for which a significant density of particles exists on the aggregate.

For the special case of normal incidence flux ($\theta = 0°$), the aggregate takes a well-defined conical structure capped by a curved dome. The special case of a zero degree flux angle has been studied by Krug and Meakin in 2+1 dimensions [72], as well as by others in 1+1 dimensions that show this characteristic shape [10, 13, 81, 83, 128, 132]. Ramanlal and Sander [132] give a first-order approximation to the curve that describes the dome on top of the aggregate,

$$r(\varphi) = r_0\sqrt{\cos\varphi}. \tag{9.1}$$

This curve fits reasonably well to the simulated profile shown in Fig. 9.9a, and also provides a theoretical prediction of the fanout angle for normal incidence [13, 179], $\varphi = \tan^{-1}\sqrt{2} = 54.73°$. To derive this fanout angle, consider the x coordinate of (9.1), $x = r\sin\varphi = r_0\sin\varphi\sqrt{\cos\varphi}$, where the definition of x is

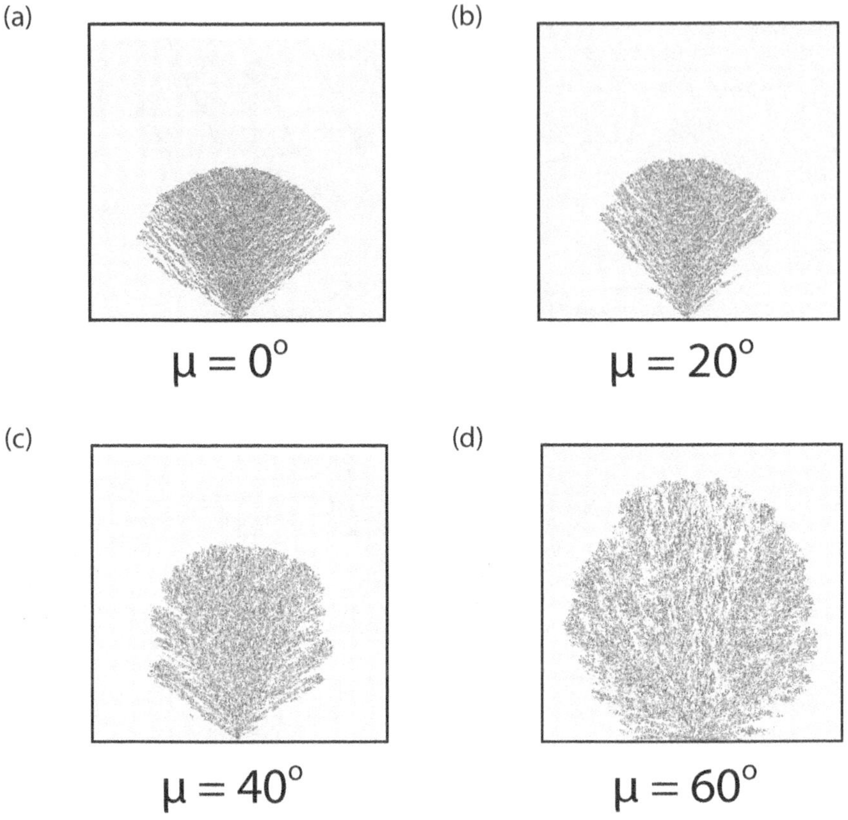

Fig. 9.9. Cross-section images of aggregate growth on a point seed with no surface diffusion. The images in (**a**) through (**d**) depict aggregate growth under oblique fluxes of angles 0°, 20°, 40°, and 60°, respectively [124].

given in Fig. 9.8. The maximum value for the x coordinate is obtained at the angle φ where $\mathrm{d}x/\mathrm{d}\varphi = 0$, or

$$\begin{aligned}\frac{\mathrm{d}x}{\mathrm{d}\varphi} &= \frac{\mathrm{d}}{\mathrm{d}\varphi}\left(r_0 \sin\varphi\sqrt{\cos\varphi}\right)\\ &= r_0\left[\cos^{3/2}\varphi - \frac{\sin^2\varphi}{2\sqrt{\cos\varphi}}\right]\\ &= r_0\cos^{3/2}\varphi\left[1 - \frac{1}{2}\tan^2\varphi\right],\end{aligned}$$

which equals zero when $\varphi = \tan^{-1}\sqrt{2}$. From Fig. 9.10, a plot of aggregate fanout angle φ versus oblique flux angle θ, the simulation result for the fanout angle at normal incidence is approximately $56° \pm 3°$, in agreement with this result.

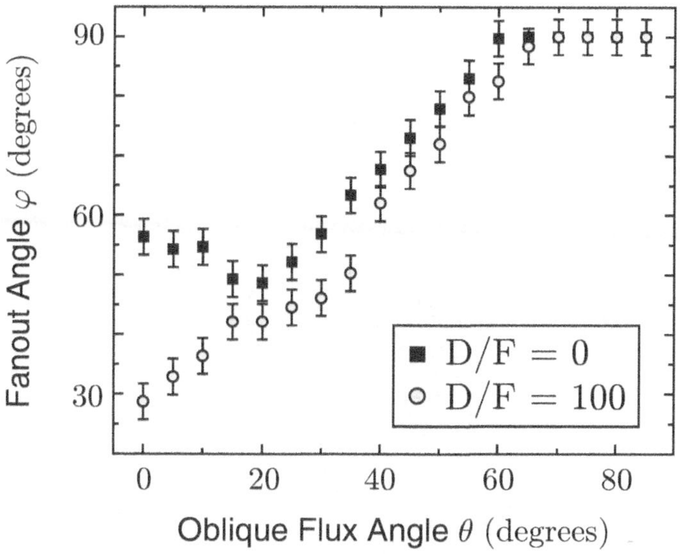

Fig. 9.10. Plot of aggregate fanout angle φ versus oblique flux angle θ with and without surface diffusion. When diffusion is inactive, a minimum fanout angle is realized at $\theta \approx 20°$ because at this oblique flux angle, the particle flux is best aligned with column growth atop the aggregate [124].

We observe an interesting behavior of the aggregate as the oblique flux angle θ is changed, as depicted in Figs. 9.9 and 9.10. As the oblique flux is increased from $\theta = 0°$ to $\theta = 20°$, the aggregate fanout angle φ decreases from approximately 56° to 48°. This behavior is clear in Fig. 9.9, as the aggregate deposited under a $\theta = 20°$ oblique flux is noticeably more compact than the aggregate formed under normal flux. However, once the oblique flux angle is increased past 20°, the fanout angle begins to increase, until at an oblique flux angle of $\theta = 60°$, there is significant growth at all angles, leading to $\varphi = 90°$.

The geometrical reasoning for this behavior is related to the presence of side-sticking in ballistic aggregation. At normal incidence, side-sticking allows the aggregate to grow laterally as well as vertically, leading to the conical structure in Fig. 9.9a. Thus, the average growth direction of a point on top of the aggregate is not parallel to the normally incident particle flux. As the oblique flux increases from 0° to 20°, the incident flux becomes more aligned with the natural growth direction of the aggregate, leading to a more compact structure.

Quantitatively, if we consider the growth from a coordinate frame at a point on top of the aggregate, it appears as if the flux were entering obliquely with respect to the local surface normal because of the characteristic dome

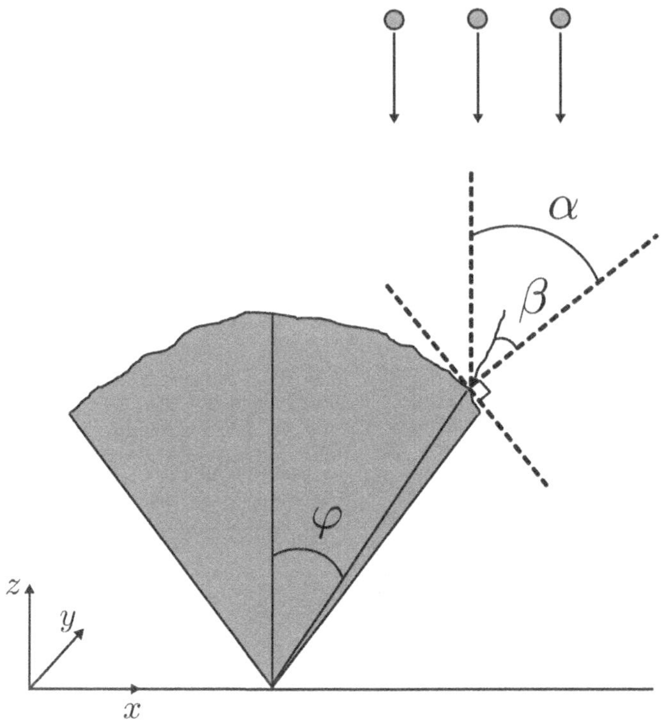

Fig. 9.11. Diagram of growth atop an aggregate for normal incidence particle flux. The particle flux makes an angle α with the local surface normal, and the local columnar structure grows at an angle β with respect to the local surface normal. It follows that the local columnar growth makes an angle $\alpha - \beta$ with the incident particle flux.

structure. From the approximate equation for the shape of the top of the aggregate at normal incidence, $r = r_0\sqrt{\cos\varphi}$, we can determine the average tilt of the local surface normal with respect to the normal particle flux. The slope of the tangent line to the curve is given by $\mathrm{d}z/\mathrm{d}x$, and the slope of the normal to the curve is given by

$$m_\perp = -\frac{1}{\mathrm{d}z/\mathrm{d}x} = -\frac{\mathrm{d}x}{\mathrm{d}z}. \tag{9.2}$$

In the x–z plane, we wish to find the angle the local surface normal makes with the z-axis, which we denote as α, because the normal flux is incident from the z direction. The definition of α is pictured in Fig. 9.11. If the slope of the local surface normal is $m_\perp$, it follows that

$$\tan\alpha = \frac{\Delta x}{\Delta z} = \frac{1}{m_\perp} = -\frac{\mathrm{d}z}{\mathrm{d}x}, \tag{9.3}$$

where $\tan\alpha = \Delta x/\Delta z$ comes from the fact that α is the angle between the local surface normal and the z-axis, not the x-axis. Therefore,

$$\alpha = \tan^{-1}\left(-\frac{\mathrm{d}z}{\mathrm{d}x}\right). \tag{9.4}$$

From the definition of the fanout angle φ in Fig. 9.8, $z = r\cos\varphi$ and $x = r\sin\varphi$ because φ is defined with reference to the z-axis ($\varphi = 0$ is the z-axis). Thus, using the chain rule,

$$\begin{aligned}
\frac{\mathrm{d}z}{\mathrm{d}x} &= \frac{\mathrm{d}z/\mathrm{d}\varphi}{\mathrm{d}x/\mathrm{d}\varphi},\\
\frac{\mathrm{d}z}{\mathrm{d}\varphi} &= \frac{\mathrm{d}r}{\mathrm{d}\varphi}\cos\varphi - r\sin\varphi\\
\frac{\mathrm{d}x}{\mathrm{d}\varphi} &= \frac{\mathrm{d}r}{\mathrm{d}\varphi}\sin\varphi + r\cos\varphi.
\end{aligned}$$

The equation of the top of the dome is $r = r_0\sqrt{\cos\varphi}$, so

$$\frac{\mathrm{d}r}{\mathrm{d}\varphi} = -\frac{r_0\sin\varphi}{2\sqrt{\cos\varphi}}.$$

Putting everything together,

$$\begin{aligned}
\alpha &= \tan^{-1}\left(-\frac{\frac{\mathrm{d}r}{\mathrm{d}\varphi}\cos\varphi - r\sin\varphi}{\frac{\mathrm{d}r}{\mathrm{d}\varphi}\sin\varphi + r\cos\varphi}\right)\\
&= \tan^{-1}\left(-\frac{-\frac{r_0\sin\varphi}{2\sqrt{\cos\varphi}}\cos\varphi - r_0\sqrt{\cos\varphi}\sin\varphi}{-\frac{r_0\sin\varphi}{2\sqrt{\cos\varphi}}\sin\varphi + r_0\sqrt{\cos\varphi}\cos\varphi}\right)\\
&= \tan^{-1}\left(\frac{3\sin\varphi\cos\varphi}{-\sin^2\varphi + 2\cos^2\varphi}\right)\\
&= \tan^{-1}\left(\frac{3\tan\varphi}{2-\tan^2\varphi}\right).
\end{aligned}$$

Therefore, at a point ($r = r_0\sqrt{\cos\varphi}, \varphi$) on the surface of the aggregate, the local surface normal makes an angle α with the normal particle flux, where

$$\alpha = \tan^{-1}\left(\frac{3\tan\varphi}{2-\tan^2\varphi}\right). \tag{9.5}$$

Taking an average along one side of the aggregate, the surface normal makes an average angle

$$\overline{\alpha} = \frac{1}{\tan^{-1}\sqrt{2}}\int_0^{\tan^{-1}\sqrt{2}} \tan^{-1}\left(\frac{3\tan\varphi}{2-\tan^2\varphi}\right)\mathrm{d}\varphi \approx 42.88^\circ$$

with the normal particle flux. If we make the assumption that, locally, the surface is flat atop the aggregate as in Fig. 9.11, the average growth atop the aggregate will mimic growth on a flat surface with a constant oblique flux of $\theta = \overline{\alpha} = 42.88°$. Using the tangent rule [132], one finds that, on average, the local columnar structure will grow at an angle of

$$\overline{\beta} = \tan^{-1}\left(\frac{1}{2}\tan\overline{\alpha}\right) \approx 24.90° \tag{9.6}$$

with respect to the local surface normal. Because the local surface normal makes an angle $\overline{\alpha}$ with the normal particle flux, the columns grow at an angle $\overline{\alpha} - \overline{\beta} = 17.98°$ with respect to the normal particle flux. (The symbols α and β in this section do not represent growth exponents as they have in the rest of the text, and the ensuing discussion utilizes the symbols p and z differently also. Unfortunately, these symbols are the adopted conventions, but their meaning should be clear from the context.)

Clearly, there are points atop the aggregate that receive flux at an angle much larger than $\overline{\alpha}$, which leads to the fanout angle of the overall aggregate to be approximately 56°. However, this argument shows that, on average, columns will grow atop the aggregate at an angle of approximately 18° from the normal particle flux, which suggests that if the incident particle flux were to be changed so as to impinge on the aggregate near this angle, the flux would tend to fall along the natural aggregate growth direction, leading to a more compact structure. This behavior is exhibited in Fig. 9.10 where at an oblique flux angle of approximately 20°, the aggregate has the least fanout. This angle is slightly higher than the predicted 18° as the expression for the shape of the top of the aggregate, $r = r_0\sqrt{\cos\varphi}$, is only valid for normal incidence and may change as the oblique flux is changed. Once the angle of the oblique flux becomes greater than 20°, we begin to observe significant lateral growth on the aggregate, often of lower density, which leads to an increase in the fanout angle. The most important conclusion to take from this discussion is that without surface diffusion, lateral growth on the aggregate becomes significant when the oblique flux angle θ is greater than 20°. This result helps explain the behavior of the aggregate shape under surface diffusion.

9.3.2 Aggregates With Diffusion

In order to make realistic predictions about the shape and density of aggregates to compare to experimental observations, we must allow for deposited particles to diffuse. The introduction of diffusion as outlined previously serves to accumulate particles inside the aggregate, as newly deposited particles will be likely to diffuse to a vacancy adjacent to a high-density volume of deposited particles. This ultimately leads to an aggregate that is more compact than those grown without surface diffusion effects. When surface diffusion effects are included, there are two competing growth effects that determine

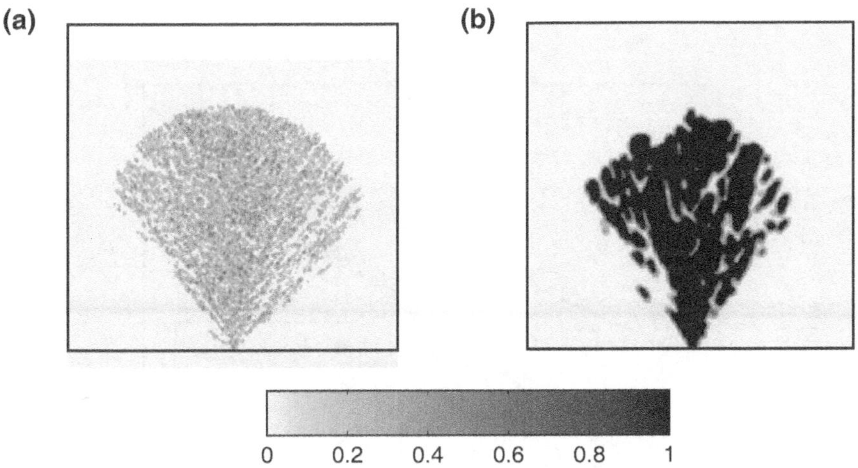

Fig. 9.12. Cross-section image of aggregate density under a 20° oblique flux (**a**) without surface diffusion, and (**b**) with surface diffusion. The profile in (**b**) shows characteristics of fingering, or small high-density growth volumes along the sides of the aggregate [124].

the density of the aggregate; the nature of the oblique flux and the diffusion. For large oblique flux angles, particles will tend to be deposited on the sides of the aggregate, which leads to low-density lateral growth. However, surface diffusion tends to move particles closer to the center of the aggregate where the density is higher. The diffusion strength in this section is taken to be $D/F = 100$ for all the results presented.

One characteristic of aggregates grown with surface diffusion is the presence of "fingering", or small high-density groups of particles on the edges of the aggregate that resemble fingers. This effect is common in experimental aggregate growth under oblique angle deposition [18, 58, 98]. Figure 9.12 includes a cross-sectional density plot of an aggregate grown under a 20° oblique flux (a) without diffusion and (b) with diffusion. Fingering is apparent in Fig. 9.12b, as well as the overall high density aggregation as compared to Fig. 9.12a.

During experimental depositions, one of the most important features to control is the average shape of the aggregate, quantified by the growth exponent p. The growth exponent p is defined as $R \sim z^p$, where R is the radius of the aggregate, and z is the height above the initial seed. Representative plots of radius R versus height z are included in Fig. 9.13. As discussed previously, normally incident flux leads to a conical aggregate, which implies a linear relationship between radius and height, or $p = 1$. This has been confirmed experimentally [179] with a fanout angle of 34.4°, which agrees with the data shown in Fig. 9.10. However, from experimental depositions with an oblique flux, a value of $p < 1$ is generally observed, with the specific value

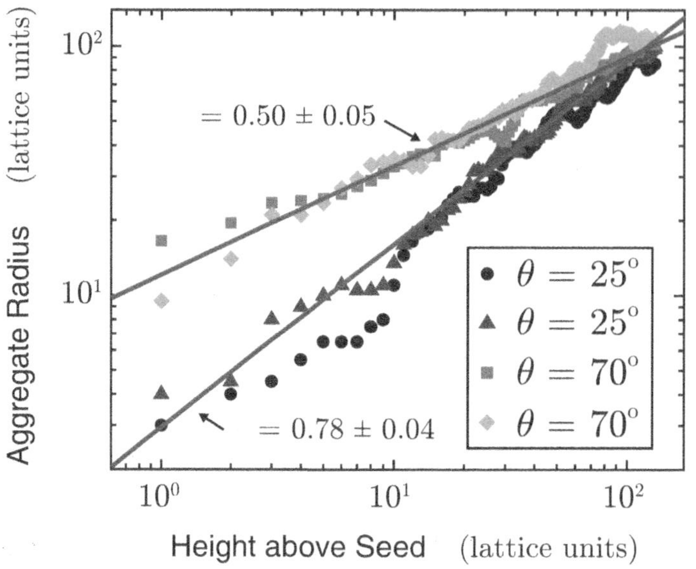

Fig. 9.13. Graph of aggregate radius R versus height above seed z used to determine the exponent p. Two simulation runs are shown for an oblique flux of $\theta = 25°$ and $\theta = 70°$ observing $p = 0.78 \pm 0.04$ and $p = 0.50 \pm 0.05$, respectively [124].

of p depending on the angle of oblique flux and strength of surface diffusion. We plot the value of the growth exponent p versus the oblique flux angle θ in Fig. 9.14. Some experimentally measured values of p reported in the literature [18, 56, 58, 98, 179] are also included in Fig. 9.14 and listed in Table 9.1.

In general, the value of p remains equal to one for oblique flux angles less than 20°, indicative of the conical structure of the aggregate. The value of p decreases as the oblique flux angle is increased, until it reaches a value of $p \approx 0.4$ for very large oblique flux angles. As observed in the simulations without surface diffusion, we see a change in the overall shape of the aggregate at an oblique flux angle of 20°, where in the case of no surface diffusion the fanout angle is a minimum from Fig. 9.10. The behavior of the growth exponent p is related to the behavior of the fanout angle discussed in Sect. 9.3.1. The fanout angle reaches a minimum in Fig. 9.10 because, at oblique flux angles larger than 20°, there is a significant amount of low-density lateral growth on the aggregate, leading to a larger fanout angle. However, with the introduction of surface diffusion, these low-density areas on the side of the aggregate diffuse towards the center of the aggregate, and change the natural linear structure of the side of the aggregate into a nonlinear shape evidenced by the smaller value of the growth exponent p. Therefore, this ballistic aggregation model predicts a value of $p < 1$ for oblique angle flux

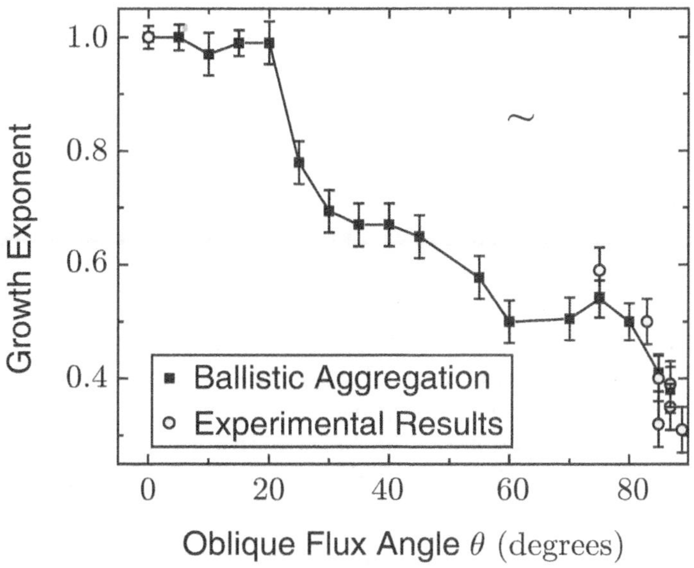

Fig. 9.14. Plot of the growth exponent p ($R \sim z^p$) versus oblique flux angle θ with surface diffusion, along with selected experimental results [18, 56, 58, 98, 179]. A value of $p = 1$ corresponds to a conical structure [124].

greater than 20°, consistent with the experimentally measured values for p shown in Fig. 9.14. In a previous work, Karabacak et al. [58] investigated the effect of surface diffusion on the growth exponent p for the special case of an oblique flux of $\theta = 85°$ through both experimental depositions and ballistic aggregation simulations without side-sticking, and found that with a similar surface diffusion strength, $p \approx 0.46$, consistent with the results for $\theta = 85°$ shown in Fig. 9.14. In particular, it was shown that as surface diffusion becomes stronger, the growth exponent p tends to decrease, which may explain the variation in experimental measurements of p from different references. As is evident from the experimental results under an 85° and 87° oblique flux, the growth exponent can vary depending on specific materials used and experimental deposition conditions. However, from Fig. 9.14, the nature of the oblique flux clearly has a strong effect on the value of the growth exponent p, and consequently on the resulting morphology of the aggregate.

In the context of the growth exponent p, we can quantify the effects of self-shadowing. Under normally incident particle flux ($\theta = 0°$), particles tend to accumulate uniformly atop the aggregate, leading to a value of $p = 1$. However, once the angular nature of the particle flux becomes significant, flux that would have accumulated atop the aggregate now accumulates along the sides of the aggregate. This then brings significance to the entire shape of the

Table 9.1. Comparison of experimentally measured growth exponents and those predicted by ballistic aggregation. Limited data are available for smaller oblique flux angles.

Oblique Flux Angle θ	Experimental p	Simulated p	Reference
$0°$	1.02 ± 0.03	1.00 ± 0.01	[179]
$75°$	0.59 ± 0.04	0.52 ± 0.03	[18]
$83°$	0.50 ± 0.04	0.45 ± 0.04	[18]
$85°$	0.40 ± 0.04	0.41 ± 0.04	[18]
	0.32 ± 0.01		[58]
$87°$	0.35 ± 0.04	0.38 ± 0.04	[18]
	0.39 ± 0.03		[98]

aggregate, not just the top of the aggregate as in normal deposition. As a result, there is a complex competition between the vertical and lateral growth velocities at every position on the aggregate, which in turn leads to $p < 1$. The deviation of p from one is then a rough measure of the significance of lateral growth on the aggregate. This lateral growth under a strong oblique flux can shadow other parts of the aggregate, which leads to self-shadowing.

We also note that in many of the reported experimental works, the radius of the aggregate very often saturates and stops growing during the later stages of growth. The aggregates discussed here do not show this saturation. A saturation of the radius R with height z indicates that no particle flux deposits on the sides of the aggregate, due to the shadowing of adjacent columns in the experimental investigations cited. As only a single column is considered when modeling a point seed, these saturation effects are not observed.

As an example of experimental results that can be described by this discussion, we show in Fig. 9.15 the growth of aggregates on a flat surface at an incident angle of 85° [58]. When films are grown on an initially flat surface under an oblique angle flux with rotation, three-dimensional islands are formed that serve as "seeds" on which aggregates can grow [58]. Due to the shadowing effect, a competition exists between the aggregates, where taller aggregates survive and shorter aggregates stop growing. We believe that the mechanism that controls the growth of aggregate size is dominated by self-shadowing. The experimental results for the growth exponent p are reasonably well explained by the ballistic aggregation model presented in this section, and a slight variation of p can be observed with different materials owing perhaps to the varying diffusion strengths of the materials.

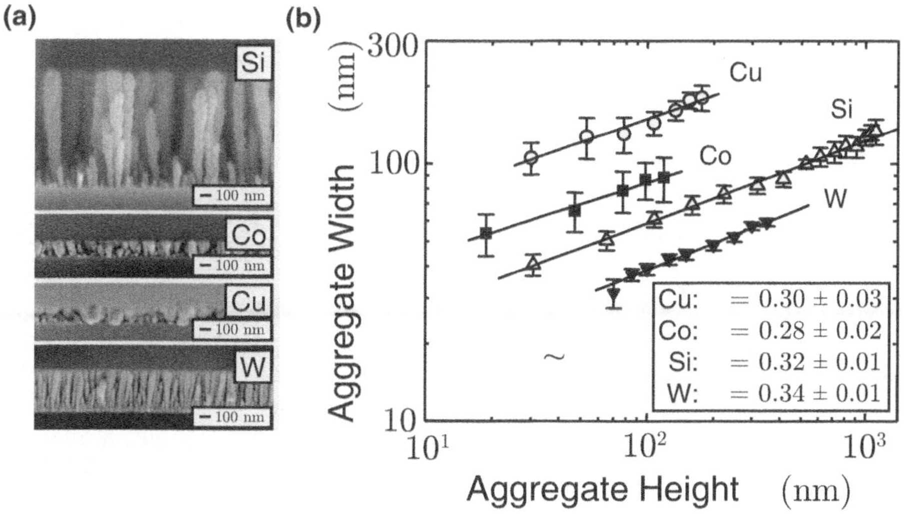

Fig. 9.15. (**a**) Cross-section view of scanning electron micrographs of Si, Co, Cu, and W aggregates grown by oblique angle deposition with substrate rotation at an 85° incident angle. (**b**) Log–log plots of the aggregate width R as a function of aggregate height z. The values of the growth exponent p are extracted from the plots are shown in the inset [58].

10

Concluding Remarks

In writing this monograph, we do not intend to provide a detailed survey of the vast number of experimental results on thin film growth reported in the literature. Instead, we hope to present a simplified view of the modeling techniques used to describe thin films grown by different experimental deposition techniques. In particular, we want to highlight the importance of nonlocal effects in thin film growth, as nonlocal effects are necessary to describe many experimental observations. For example, Fig. 4.1 provides a summary of the range of growth exponents β predicted by both local and nonlocal models. Local models, both continuum models and discrete models, typically give a β value smaller than 0.25. On the other hand, depending on the magnitude of the sticking coefficient s_0, the competition between shadowing and reemission in nonlocal models can give a β value in the range 0–1. In the same graph, we also display a characteristic range of β values reported in experimental papers in the literature for different deposition techniques including thermal evaporation, sputter deposition, chemical vapor deposition, and oblique angle deposition. It appears that local models account for the measured exponent values reasonably well for thermal evaporation with normally incident flux, a situation where there is no geometrical shadowing. The sticking coefficient for most materials in thermal evaporation is equal to 1 because particles in the incident flux tend to have a small kinetic energy, typically less than a fraction of an eV. In this case, dynamic scaling should be applicable, and the surface is likely to be self-affine with no mound formation.

For sputter deposition, the majority of the experimental values of β lie somewhere in the narrow range of 0.4–0.5. This translates into a sticking coefficient in the range of 0.5–0.7 in a reemission model, which is believed to be the case for the silicon family of materials, which comprise the majority of reported experimental results [59, 122, 137]. During sputtering, atoms tend to be energetic with a kinetic energy of several eV, which would lead to a sticking coefficient significantly less than 1. In this case, the surface is no longer self-affine and mounds are formed through shadowing, which leads to the breakdown of dynamic scaling. For chemical vapor deposition, the experimental value of β reported in the literature is more spread out. For CVD, the sticking coefficient can vary significantly depending on the materials used and the temperature at which the materials are deposited.

Mound formation tends to dominate when the sticking coefficient s_0 is larger than 0.5. The growth of the separation of mounds λ for both sputter deposition and chemical vapor deposition in the large sticking coefficient regime exhibits a power-law behavior, $\lambda \sim t^p$. There is a strong indication that the exponent associated with this length scale is $p \approx 0.5$, and is universal. It would be interesting to investigate this prediction in future experiments utilizing a diverse set of materials. For oblique angle deposition, severe shadowing occurs to give a variety of isolated nanostructures where the β value appears to be always 1. In this case, wavelength selection also gives $p \approx 0.5$ [123, 156].

Although most of these predictions are based on the results of solid-on-solid models because of their simplicity, models including overhang structures may be more realistic when films grow thicker than a few microns. Due to the challenge in defining the surface profile for a surface with overhangs, as is observed in a ballistic aggregation model, fewer attempts to quantify such a surface profile and compare results to experimental works have been reported in the literature. For a thicker film, an overhang structure can effectively serve as a smoothing mechanism, similar to the effect of a nonunity sticking coefficient. Research in this direction would be of interest. It may even shine some light on a possible quantitative description of the structural zone models [159] that have often been used to qualitatively describe sputter deposited films that are typically tens of microns thick, especially in Zone I, the low-temperature regime where diffusion is not excessive.

Most of the discussions in this monograph have focused on growth that occurs far-from-equilibrium. However, for many realistic systems and especially chemical vapor deposition, the deposition may occur at a sufficiently high temperature that the system is not far-from-equilibrium and microscopic energetics may come into play. A common growth phenomenon controlled by energetics is the process of nucleation and growth, where the surface energy becomes important. The surface energy competition between the substrate and depositing materials leads to the well-known three modes of film growth [163]: layer-by-layer growth (Frank–Van der Merwe growth mode), island growth (Volmer–Weber growth mode), and intermediate growth (Stranski–Krastanov growth mode). These are of course a separate area of research.

We hope that the discussions and examples provided in this monograph have given the reader a comprehensive overview of thin film growth modeling, and aid in the many new discoveries waiting to be found in the field.

A

Mathematical Appendix

The mathematics presented in this appendix is intended to supplement the discussion and formulas given in the main body of the text. The reader is assumed to be familiar with elementary calculus and complex variables, although a brief discussion of complex variables is given when appropriate. The goal of this appendix is to avoid any confusion in understanding the mathematics presented in the main text, and as such derivations and discussions are drawn out in significant detail. In particular, the techniques introduced are necessary for taking the material presented in the text one step further and experimenting with one's own mathematical models for thin film morphology. Mathematical rigor is not stressed in this discussion; for a more complete derivation and discussion of the following topics, see [26, 45, 102].

A.1 Special Functions

In the mathematical description of thin film morphology, a number of special functions are used that may be unfamiliar to the reader, and it is useful to describe their behavior and properties. In particular, a class of functions called Bessel functions are significantly utilized. We begin with the Bessel function of the first kind.

A.1.1 Bessel Function of the First Kind

The nth order Bessel function of the first kind, denoted $J_n(x)$, is defined as

$$J_n(x) = \sum_{j=0}^{\infty} \frac{(-1)^j}{j!\Gamma(n+j+1)} \left(\frac{x}{2}\right)^{2j+n}, \tag{A.1}$$

where $\Gamma(x)$ is the gamma function discussed in Sect. A.1.4. In particular, for $n = 0$,

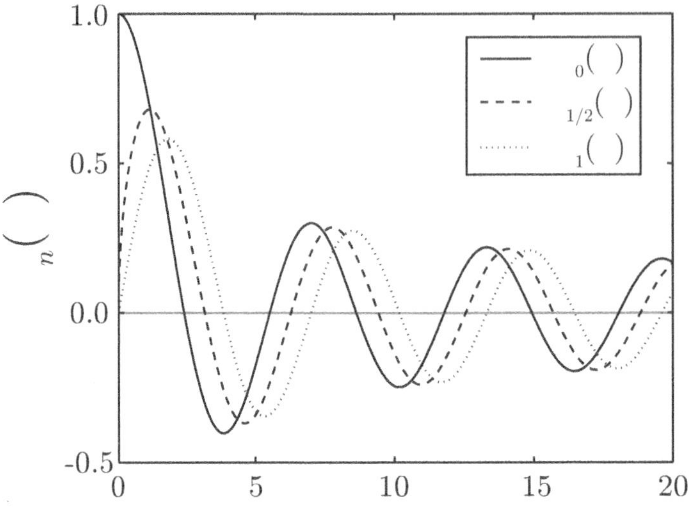

Fig. A.1. Ordinary Bessel functions $J_n(x)$ of order $n = 0$, $1/2$, and 1.

$$J_0(x) = \sum_{j=0}^{\infty} \frac{(-1)^j}{(j!)^2} \left(\frac{x}{2}\right)^{2j} . \tag{A.2}$$

Bessel functions have a characteristic oscillatory behavior, and for asymptotically large x behave as cos functions,

$$J_n(x) \sim \sqrt{\frac{2}{\pi x}} \cos\left(x - \frac{n\pi}{2} - \frac{\pi}{4}\right), \quad x \gg \left|n^2 - \frac{1}{4}\right| . \tag{A.3}$$

This oscillatory behavior is clear from Fig. A.1. Similarly for small x, from the power series in (A.1),

$$J_n(x) \sim \frac{1}{n!} \left(\frac{x}{2}\right)^n , \quad x \ll \sqrt{n+1}. \tag{A.4}$$

An important relation involving Bessel functions is the orthogonality relation on a semi-infinite interval,

$$\int_0^\infty x J_n(ax) J_n(bx) \mathrm{d}x = \frac{1}{a} \delta(a-b), \tag{A.5}$$

where $\delta(x)$ is the delta function, described in Sect. A.1.5, which implies that the integral is zero unless $a = b$. This property of Bessel functions is a key reason why they are incorporated into models for 2+1 dimensional surfaces because this orthogonality leads to analytical expressions for surface statistics such as the power spectral density function.

Another formula that is used frequently is the relation

$$\int_0^{2\pi} \mathrm{e}^{\mathrm{i}x\cos\theta}\mathrm{d}\theta = 2\pi J_0(x). \tag{A.6}$$

This can be derived by expanding the exponential in a power series,

$$\int_0^{2\pi} \sum_{n=0}^{\infty} \frac{(\mathrm{i}x\cos\theta)^n}{n!}\mathrm{d}\theta = \sum_{n=0}^{\infty} \frac{(\mathrm{i}x)^n}{n!} \int_0^{2\pi} \cos^n\theta\mathrm{d}\theta. \tag{A.7}$$

Using complex variable techniques as outlined in Sect. A.2, it can be shown that

$$\int_0^{2\pi} \cos^n\theta\mathrm{d}\theta = \begin{cases} 0, & n \text{ odd}, \\ \frac{2\pi}{2^n}\frac{n!}{[(n/2)!]^2}, & n \text{ even}. \end{cases} \tag{A.8}$$

This result gives, with $n = 2j$,

$$\begin{aligned} \sum_{n=0}^{\infty} \frac{(\mathrm{i}x)^n}{n!} \int_0^{2\pi} \cos^n\theta\mathrm{d}\theta &= \sum_{j=0}^{\infty} \frac{(\mathrm{i}x)^{2j}}{(2j)!}\frac{2\pi}{2^{2j}}\frac{(2j)!}{(j!)^2} \\ &= 2\pi \sum_{j=0}^{\infty} \frac{(-1)^j}{(j!)^2}\left(\frac{x}{2}\right)^{2j} \\ &= 2\pi J_0(x), \end{aligned}$$

using the definition of $J_0(x)$ from (A.2).

A.1.2 Modified Bessel Function of the First Kind

The modified Bessel function of the first kind, denoted by $I_n(x)$, is a notation used to denote the ordinary Bessel function of the first kind for a purely complex argument. The modified class of functions is defined as

$$I_n(x) = \mathrm{i}^{-n} J_n(\mathrm{i}x). \tag{A.9}$$

These functions have an exponential behavior for large x,

$$I_n(x) \sim \frac{\mathrm{e}^x}{\sqrt{2\pi x}}, \quad x \gg \left| n^2 - \frac{1}{4} \right|. \tag{A.10}$$

The most useful relation for our purposes involving the modified Bessel function of the first kind is obtained from (A.6) with the substitution $t = \mathrm{i}x$,

$$\int_0^{2\pi} \mathrm{e}^{t\cos\theta}\mathrm{d}\theta = 2\pi I_0(t). \tag{A.11}$$

A.1.3 Modified Bessel Function of the Second Kind

The modified Bessel function of the second kind, denoted $K_n(x)$, is defined as

$$K_n(x) = \frac{\pi}{2}\frac{I_{-n}(x) - I_n(x)}{\sin n\pi}. \tag{A.12}$$

This expression is indeterminate for integer n, and an appropriate limit must be taken for integer n. The limiting behavior of this function is, for large x,

$$K_n(x) \sim \sqrt{\frac{\pi}{2x}}\mathrm{e}^{-x}, \quad x \gg \left|n^2 - \frac{1}{4}\right|, \tag{A.13}$$

and for small x and $|n| < 1$, using (A.1),

$$K_n(x) \sim \frac{\Gamma(n)}{2}\left(\frac{2}{x}\right)^n - \frac{\Gamma(1-n)}{2n}\left(\frac{x}{2}\right)^n, \quad x \ll \sqrt{n+1}, \tag{A.14}$$

where $\Gamma(x)$ is the gamma function described in Sect. A.1.4. Two specific cases are of interest, the behavior of the function for orders 0 and 1/2, given by [4, 15]:

$$K_{1/2}(x) = \sqrt{\frac{\pi}{2x}}\mathrm{e}^{-x}, \tag{A.15}$$

$$K_0(x) \sim -\ln x. \tag{A.16}$$

A.1.4 Gamma Function

The gamma function $\Gamma(x)$ is defined as

$$\Gamma(x) = \int_0^\infty s^{x-1}\mathrm{e}^{-s}\mathrm{d}s. \tag{A.17}$$

This function is most commonly known for providing a continuous generalization of the factorial function $x! = x(x-1)(x-2)\cdots 2\cdot 1$, as can be shown by an integration by parts,

$$\begin{aligned}
\Gamma(x) &= \int_0^\infty s^{x-1}\mathrm{e}^{-s}\mathrm{d}s \\
&= \left[-s^{x-1}\mathrm{e}^{-s}\right]_0^\infty + \int_0^\infty (x-1)s^{x-2}\mathrm{e}^{-s}\mathrm{d}s \\
&= (x-1)\int_0^\infty s^{x-2}\mathrm{e}^{-s}\mathrm{d}s \\
&= (x-1)\Gamma(x-1).
\end{aligned}$$

Using this property of $\Gamma(x)$, coupled with the value of $\Gamma(1) = 1$, it follows by induction that if x is an integer, then

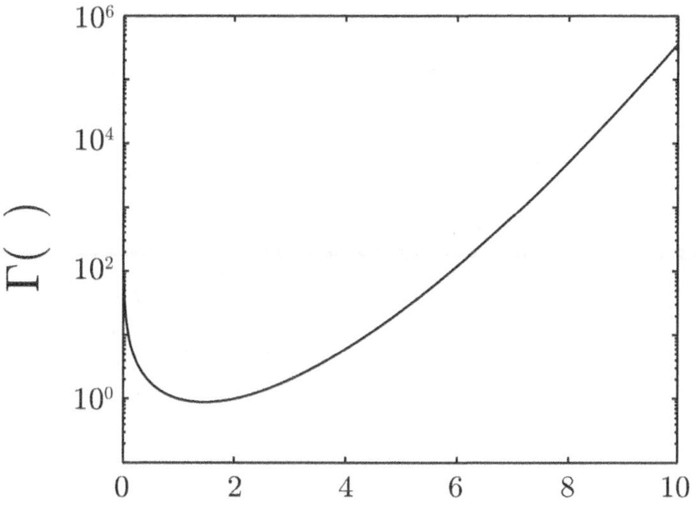

Fig. A.2. Gamma function $\Gamma(x)$ for $0 < x \leq 10$. For integer x, $\Gamma(x) = (x-1)!$.

$$\Gamma(x) = (x-1)!. \tag{A.18}$$

However, whereas the factorial function is only defined for positive integers, $\Gamma(x)$ is continuously defined for $x > 0$. A graph of $\Gamma(x)$ is provided in Fig. A.2. Another useful property of $\Gamma(x)$ is

$$\Gamma(x)\Gamma(1-x) = \frac{\pi}{\sin x\pi}, \tag{A.19}$$

which was used to find the series expansion of the modified Bessel function of the second kind in the previous section.

A.1.5 Delta Function

The delta function $\delta(x)$ can be thought of conceptually as an infinitely short pulse. For most physical purposes, the delta function can be defined as

$$\delta(x) = \begin{cases} \infty, & x = 0, \\ 0, & x \neq 0, \end{cases} \tag{A.20}$$

with the additional property that

$$\int_{-\infty}^{\infty} \delta(x)\mathrm{d}x = 1. \tag{A.21}$$

However, the traditional Riemann integral of any well-behaved function that is zero everywhere except at one point is zero, which contradicts (A.21). This contradiction arises because $\delta(x)$ is not a true function, and a rigorous mathematical definition of the delta function would be beyond the scope of this discussion. Even though $\delta(x)$ is not a function itself, it can be thought of as the limit of a sequence of functions whose properties approach the properties of the delta function. For example, consider the function of x defined by the limit

$$f(x) = \lim_{\eta \to 0^+} \frac{1}{\pi} \frac{\eta}{x^2 + \eta^2}, \tag{A.22}$$

where $\lim_{\eta \to 0^+}$ means that the limit is approaching zero from positive η. This can be thought of as a sequence of functions of x that possesses all the properties of a delta function. For example, the integral of each member of the sequence is given by

$$\begin{aligned} \int_{-\infty}^{\infty} \frac{1}{\pi} \frac{\eta}{x^2 + \eta^2} \mathrm{d}x &= \frac{1}{\pi} \left[\tan^{-1} \frac{x}{\eta} \right]_{-\infty}^{\infty} \\ &= \frac{1}{\pi} \left[\frac{\pi}{2} + \frac{\pi}{2} \right] \\ &= 1, \end{aligned}$$

which is independent of η. Also, for $x \neq 0$,

$$\lim_{\eta \to 0^+} \frac{1}{\pi} \frac{\eta}{x^2 + \eta^2} = 0, \tag{A.23}$$

and, for $x = 0$,

$$\lim_{\eta \to 0^+} \frac{1}{\pi} \frac{\eta}{0^2 + \eta^2} = \lim_{\eta \to 0^+} \frac{1}{\pi \eta} = \infty. \tag{A.24}$$

Thus, if we make the designation

$$\delta(x) = \lim_{\eta \to 0^+} \frac{1}{\pi} \frac{\eta}{x^2 + \eta^2}, \tag{A.25}$$

the properties in (A.20) and (A.21) are satisfied. There are many such ways to define the delta function as the limit of a series of functions. The most useful definition for our purposes is the relation

$$\delta(x) = \frac{1}{2\pi} \int_{-\infty}^{\infty} \mathrm{e}^{\mathrm{i}kx} \mathrm{d}k, \tag{A.26}$$

or, more specifically, the limit of the following sequence with integer n,

$$\delta(x) = \lim_{n \to \infty} \frac{1}{2\pi} \int_{-n\pi/x}^{n\pi/x} \mathrm{e}^{\mathrm{i}kx} \mathrm{d}k. \tag{A.27}$$

Let us verify that this is indeed a valid definition of the delta function. We can perform the integration with $x \neq 0$ to obtain

$$\begin{aligned}\int_{-n\pi/x}^{n\pi/x} \mathrm{e}^{\mathrm{i}kx}\mathrm{d}k &= \left[\frac{1}{\mathrm{i}x}\mathrm{e}^{\mathrm{i}kx}\right]_{-n\pi/x}^{n\pi/x} \\ &= \left[\frac{1}{\mathrm{i}x}\left(\mathrm{e}^{\mathrm{i}n\pi} - \mathrm{e}^{-\mathrm{i}n\pi}\right)\right] \\ &= \frac{2\sin n\pi}{x}.\end{aligned}$$

In (A.27), n is an integer, so the value of this integral is zero for each n when $x \neq 0$. Clearly, when $x = 0$, the integral in (A.27) becomes

$$\int_{-\infty}^{\infty} \mathrm{e}^0 \mathrm{d}k = \infty. \tag{A.28}$$

Thus, the limit in (A.27) satisfies the properties of (A.20). Now we must show that (A.21) holds as well. To do so, consider the general case of the limits of integration in (A.27) to be any number a, and examine the limit as $a \to \infty$.

$$\lim_{a\to\infty} \frac{1}{2\pi} \int_{-\infty}^{\infty} \int_{-a}^{a} \mathrm{e}^{\mathrm{i}kx}\mathrm{d}k\mathrm{d}x = \lim_{a\to\infty} \frac{1}{2\pi} \int_{-\infty}^{\infty} \frac{2\sin ax}{x}\mathrm{d}x. \tag{A.29}$$

The variable a was used instead of $n\pi/x$ because the special case of $n\pi/x$ leads to a question of convergence of the integral; in the end the result will not depend on a and will be valid for any a. This is the price that is paid for avoiding a mathematically rigorous derivation because interchanging the orders of integration is not necessarily justified, but the results we obtain will be correct. If we change variables in (A.29) to $t = ax$, the integral then becomes

$$\lim_{a\to\infty} \frac{1}{2\pi} \int_{-\infty}^{\infty} \frac{2\sin ax}{x}\mathrm{d}x = \lim_{a\to\infty} \frac{1}{\pi} \int_{-\infty}^{\infty} \frac{\sin t}{t}\mathrm{d}t, \tag{A.30}$$

which is independent of a. Using complex variable techniques, the value of the last integral can be shown to be

$$\int_{-\infty}^{\infty} \frac{\sin t}{t}\mathrm{d}t = \pi, \tag{A.31}$$

so the integral of each member of the sequence in (A.27) is one, and (A.21) is satisfied.

The utility of the delta function lies in what is commonly called a "sifting" property,

$$\int_{-\infty}^{\infty} f(x)\delta(x)\mathrm{d}x = f(0) \tag{A.32}$$

for a well-behaved function $f(x)$ or, more generally,

$$\int_{-\infty}^{\infty} f(x)\delta(x-a)\mathrm{d}x = f(a). \tag{A.33}$$

Because $\delta(x-a)=0$ for $x \neq a$, the only point in the domain of integration that matters is $x=a$. In addition, the delta function is normalized to one from (A.21), therefore the result of the integral is simply the function value at $x=a$. Thus, when the delta function appears under an integral, it "picks out" the value of the function it multiplies and renders the integral trivial to evaluate.

The previous derivations have been carried out in one dimension, but the delta function can be defined in multiple dimensions. For example, the two-dimensional delta function can be written as

$$\delta^2(\mathbf{x}) = \delta(x)\delta(y), \tag{A.34}$$

where $\mathbf{x}=(x,y)$ and the symbol δ^2 is used to denote that the delta function is two-dimensional; it does not mean that the delta function is squared. In general,

$$\delta^d(\mathbf{x}) = \prod_{j=1}^{d} \delta(x_d), \tag{A.35}$$

where $\mathbf{x}$ is a d-dimensional vector, and x_d are Cartesian coordinates. It should also be noted that these definitions only hold in Cartesian coordinates, for example, in two-dimensional polar coordinates [16],

$$\delta^2(\mathbf{x}) = \frac{\delta(r)}{\pi|r|}. \tag{A.36}$$

A.2 Complex Integrals

For the derivations that are included in the following sections, certain concepts in complex variables are required to carry out the mathematics. For readers unfamiliar with complex numbers, we present a short discussion of the concepts needed to follow the mathematics that ensues.

Complex numbers can be represented in terms of real numbers along with the symbol i, which is defined as $\mathrm{i}^2=-1$. A complex number z can be written in terms of the real numbers x and y as

$$z = x + \mathrm{i}y. \tag{A.37}$$

Carrying out operations with complex numbers follows the same rules as those with real numbers, assuming that whenever i^2 appears, it is replaced with -1. One of the most well-known formulas in complex numbers is Euler's formula,

$$\mathrm{e}^{\mathrm{i}x} = \cos x + \mathrm{i}\sin x, \tag{A.38}$$

which can be shown by expanding the exponential in a Taylor series,

$$\begin{aligned} e^{ix} &= \sum_{n=0}^{\infty} \frac{(ix)^n}{n!} \\ &= \sum_{j=0}^{\infty} \frac{(ix)^{2j}}{(2j)!} + \sum_{j=0}^{\infty} \frac{(ix)^{2j+1}}{(2j+1)!} \\ &= \sum_{j=0}^{\infty} \frac{(-1)^j x^{2j}}{(2j)!} + i\sum_{j=0}^{\infty} \frac{(-1)^j x^{2j+1}}{(2j+1)!} \\ &= \cos x + i \sin x. \end{aligned}$$

Integration of a function of a complex variable is defined in a similar manner to integration in real variables, except for the specification of a path in complex variables. In ordinary single-variable calculus, the integral is taken along the x-axis, but in complex variables, an integral can be taken along any path in the x–y plane. Integration of a complex function is of interest here because it allows for the evaluation of real-valued integrals that are difficult to evaluate without complex numbers. We are interested in the specific case of an integral over a closed path, or contour, in the complex plane. If the function $f(z)$ has no singularities inside a contour C and is a function of z only and not its complex conjugate, then

$$\oint_C f(z)dz = 0. \tag{A.39}$$

Such a function is called an analytic function in the domain enclosed by C. If $f(z)$ has a singularity at $z = z_0$ of the form

$$f(z) = \frac{g(z)}{(z - z_0)^{n+1}}, \tag{A.40}$$

where $g(z)$ is a well-behaved function with no singularities in the contour, then

$$\oint_C f(z)dz = \frac{2\pi i}{n!} \lim_{z \to z_0} \frac{d^n}{dz^n} g(z). \tag{A.41}$$

This is known as Cauchy's residue theorem, and its proof is beyond the scope of our discussion. The residue of $f(z)$ is defined as

$$\mathrm{Res} f(z_0) = \frac{1}{n!} \lim_{z \to z_0} \frac{d^n}{dz^n} g(z). \tag{A.42}$$

The utility of this result lies in expressing the contour integral in terms of a real-valued integral. As an example, we derive a result used to prove (A.8). Consider the contour integral over a semicircular contour of radius R in the upper half-plane of the integrand $(1 + z^2)^{-(n+1)}$,

$$\oint \frac{dz}{(1 + z^2)^{n+1}} = \int_{-R}^{R} \frac{dx}{(1 + x^2)^{n+1}} + \int_{C_S} \frac{dz}{(1 + z^2)^{n+1}}.$$

The integral over the semicircle $\left(z = Re^{i\theta}\right)$ behaves as

$$\int_{C_S} \frac{dz}{(1+z^2)^{n+1}} = \int_0^{\pi} \frac{iRe^{i\theta}d\theta}{(1+R^2e^{2i\theta})^{n+1}} \sim O(R^{-2n-1}),$$

and approaches zero for $n \geq 1$ as $R \to \infty$. The contour integral has one singularity in the domain of integration at $z = i$ of order $n+1$. The residue at this point is given by

$$\begin{aligned}
\text{Res} f(i) &= \lim_{z\to i} \frac{1}{n!}\frac{d^n}{dz^n}\left[\frac{1}{(z+i)^{n+1}}\right] \\
&= \lim_{z\to i} \frac{1}{n!}\left[\frac{(n+1)(n+2)\cdots(2n)(-1)^n}{(z+i)^{2n+1}}\right] \\
&= \frac{1}{n!}\left[\frac{(n+1)(n+2)\cdots(2n)(i^2)^n}{2^{2n+1}i^{2n+1}}\right] \\
&= \frac{1}{n!}\left[\frac{(2n)!}{n!}\frac{1}{2^{2n+1}i}\right] \\
&= \frac{1}{2^{2n+1}i}\frac{(2n)!}{(n!)^2}.
\end{aligned}$$

By Cauchy's theorem,

$$\oint \frac{dz}{(1+z^2)^{n+1}} = 2\pi i\left[\frac{1}{2^{2n+1}i}\frac{(2n)!}{(n!)^2}\right] = \frac{\pi}{2^{2n}}\frac{(2n)!}{(n!)^2}.$$

It follows that

$$\int_{-\infty}^{\infty} \frac{dx}{(1+x^2)^{n+1}} = \frac{\pi}{2^{2n}}\frac{(2n)!}{(n!)^2}.$$

Now, make the transformation $x = \tan\theta$, $dx = \sec^2\theta d\theta$,

$$\begin{aligned}
\int_{-\infty}^{\infty} \frac{dx}{(1+x^2)^{n+1}} &= \int_{-\pi/2}^{\pi/2} \frac{\sec^2\theta d\theta}{(1+\tan^2\theta)^{n+1}} \\
&= \int_{-\pi/2}^{\pi/2} \frac{\sec^2\theta d\theta}{\sec^{2n+2}\theta} \\
&= \int_{-\pi/2}^{\pi/2} \cos^{2n}\theta d\theta.
\end{aligned}$$

Because $\cos^2\theta$ is periodic in π, the integral over the interval $[-\pi/2, \pi/2]$ is an integral over one period, so shifting the domain to $[0, \pi]$ will not affect the integral because the integral is still over one full period. Extending the domain to $[0, 2\pi]$ will double the value of the integral because the domain $[0, 2\pi]$ is over two full periods, which gives

$$\int_0^{2\pi} \cos^{2n}\theta d\theta = 2\int_0^{\pi} \cos^{2n}\theta d\theta = \frac{2\pi}{2^{2n}}\frac{(2n)!}{(n!)^2}. \tag{A.43}$$

This complex variable contour integration technique is very powerful, and can be used to show the following two results that are utilized in this appendix,

$$\int_{-\infty}^{\infty} \frac{\sin x}{x} \mathrm{d}x = \pi$$

$$\int_0^{2\pi} \frac{\mathrm{d}\theta}{(1 - a\cos\theta)^2} = \frac{2\pi}{(1-a^2)^{3/2}}.$$

These results are left to the reader to verify.

A.3 Fourier Transform of a Product

The Fourier transform of a function is the representation of a function in frequency space. The set of functions $\{\mathrm{e}^{\mathrm{i}kx}\}$, for real k, constitutes a complete set of functions, therefore any arbitrary function $f(x)$ can be decomposed into a sum of these functions. Recall that $\mathrm{e}^{\mathrm{i}kx} = \cos kx + \mathrm{i}\sin kx$, so this statement suggests that $f(x)$ can be represented by a weighted sum of waves of different wavenumbers k. If $f(x)$ is not periodic, an integral over all real k will be required to represent $f(x)$ in frequency space. However, if $f(x)$ is periodic, only the harmonic frequencies corresponding to the period of $f(x)$ will contribute, which is called a Fourier series because the sum is discrete as opposed to an integral. This discussion suggests that any function $f(x)$ can be represented as

$$f(x) = \int_{-\infty}^{\infty} \hat{f}(k)\mathrm{e}^{-\mathrm{i}kx}\mathrm{d}k, \tag{A.44}$$

where $\hat{f}(k)$ is the Fourier transform of $f(x)$. The integral is simply a sum over all frequencies, or wavenumbers k, with an appropriate weight for each wavenumber, given by $\hat{f}(k)$. In this section, unless otherwise noted, all integrals are taken from $-\infty$ to ∞. Multiplying both sides of this equation by $\mathrm{e}^{\mathrm{i}k'x}$ and integrating over x gives

$$\int f(x)\mathrm{e}^{\mathrm{i}k'x}\mathrm{d}x = \int\int \hat{f}(k)\mathrm{e}^{\mathrm{i}(k'-k)x}\mathrm{d}k\mathrm{d}x. \tag{A.45}$$

However, (A.26) can be used to perform the integral over x as a delta function,

$$\int f(x)\mathrm{e}^{\mathrm{i}k'x}\mathrm{d}x = 2\pi \int \hat{f}(k)\delta(k'-k)\mathrm{d}k$$

$$\frac{1}{2\pi}\int f(x)\mathrm{e}^{\mathrm{i}k'x}\mathrm{d}x = \hat{f}(k').$$

Therefore, we can define the Fourier transform of $f(x)$, denoted by $\hat{f}(k)$ or $\mathcal{F}\{f(x)\}$, as

$$\mathcal{F}\{f(x)\} = \hat{f}(k) = \frac{1}{2\pi}\int f(x)\mathrm{e}^{\mathrm{i}kx}\mathrm{d}x. \tag{A.46}$$

Motivated by this definition in one dimension, in d-dimensions the Fourier transform becomes

$$\mathcal{F}\{f(\mathbf{x})\} = \frac{1}{(2\pi)^d}\int f(\mathbf{x})\mathrm{e}^{\mathrm{i}\mathbf{k}\cdot\mathbf{x}}\mathrm{d}\mathbf{x}. \tag{A.47}$$

Now, consider the Fourier transform of the product of two functions $f(x)g(x)$. This can be written as, using (A.26),

$$\begin{aligned}
\mathcal{F}\{f(x)g(x)\} &= \frac{1}{2\pi}\int f(x)g(x)\mathrm{e}^{\mathrm{i}kx}\mathrm{d}x \\
&= \frac{1}{2\pi}\int\int f(x)g(x')\delta(x'-x)\mathrm{e}^{\mathrm{i}kx}\mathrm{d}x\mathrm{d}x' \\
&= \frac{1}{(2\pi)^2}\int\int\int f(x)g(x')\mathrm{e}^{\mathrm{i}k'(x'-x)}\mathrm{e}^{\mathrm{i}kx}\mathrm{d}x\mathrm{d}x'\mathrm{d}k' \\
&= \int \mathrm{d}k'\left(\frac{1}{2\pi}\int \mathrm{d}x f(x)\mathrm{e}^{\mathrm{i}(k-k')x}\right)\left(\frac{1}{2\pi}\int \mathrm{d}x' g(x')\mathrm{e}^{\mathrm{i}k'x'}\right).
\end{aligned}$$

Let us define the convolution $(F{*}G)(k)$ of two functions $F(k)$ and $G(k)$ as

$$(F{*}G)(k) = \int \mathrm{d}k' F(k-k')G(k'). \tag{A.48}$$

It then follows that

$$\mathcal{F}\{f(x)g(x)\} = \mathcal{F}\{f(x)\}*\mathcal{F}\{g(x)\}; \tag{A.49}$$

that is, the Fourier transform of the product of two functions is the convolution of the individual Fourier transforms. Note that the convolution was defined in terms of Cartesian coordinates, and in higher dimensions, Cartesian coordinates must be used to express the convolution.

A specific case of a Fourier transform that is useful is the two-dimensional Fourier transform of a product of two functions of r in polar coordinates. The Fourier transform of the product will simply be the convolution of the individual Fourier transforms, but the form can be simplified further because both functions are only functions of r.

$$\begin{aligned}
\mathcal{F}\{f(r)g(r)\} &= \mathcal{F}\{f(r)\}*\mathcal{F}\{g(r)\} \\
&= \left(\frac{1}{(2\pi)^2}\int_0^{2\pi}\int_0^{\infty} f(r)\mathrm{e}^{ikr\cos\theta}r\mathrm{d}r\mathrm{d}\theta\right) \\
&\quad * \left(\frac{1}{(2\pi)^2}\int_0^{2\pi}\int_0^{\infty} g(r)\mathrm{e}^{ikr\cos\theta}r\mathrm{d}r\mathrm{d}\theta\right).
\end{aligned}$$

Using (A.6), the integration over θ can be carried out because the functions f and g do not depend on θ,

$$\mathcal{F}\{f(r)g(r)\} = \left(\frac{1}{2\pi}\int_0^\infty f(r)J_0(kr)r\mathrm{d}r\right) * \left(\frac{1}{2\pi}\int_0^\infty g(r)J_0(kr)r\mathrm{d}r\right). \tag{A.50}$$

This form invites the definition of the Hankel transform $\mathcal{H}$,

$$\mathcal{H}\{f(r)\} = \int_0^\infty f(r)J_0(kr)r\mathrm{d}r, \tag{A.51}$$

and the Fourier transform of $f(r)g(r)$ can be written as

$$\mathcal{F}\{f(r)g(r)\} = \frac{1}{(2\pi)^2}\mathcal{H}\{f(r)\} * \mathcal{H}\{g(r)\}. \tag{A.52}$$

If the autocorrelation function $R(r)$ for a surface is the product of two functions, $R(r) = f(r)g(r)$, then the power spectral density function of the surface, which is equal to the interface width squared (w^2) times the Fourier transform of the autocorrelation function from (2.16), can be written as

$$P(k) = \frac{w^2}{(2\pi)^2}\mathcal{H}\{f(r)\} * \mathcal{H}\{g(r)\}. \tag{A.53}$$

This result is used extensively in the following sections. When using this formula, the convolution of two functions $F(r)$ and $G(r)$ expressed in terms of the polar coordinate $r = \sqrt{x^2 + y^2}$ must be written as

$$F(r)*G(r) = \int\int F\left(\sqrt{(x-x')^2 + (y-y')^2}\right) G\left(\sqrt{(x')^2 + (y')^2}\right) \mathrm{d}x'\mathrm{d}y', \tag{A.54}$$

where the integration is over the entire domain of x' and y'. The resulting integral can then be converted to a coordinate system where it is conveniently evaluated.

A.4 Power Spectral Density Functions

The mathematical details of finding the analytic forms of power spectral density functions given in the text are included in the following sections. The mathematics are carried out for the PSD in 2+1 dimensions, but similar steps may be taken to find the form in 1+1 dimensions. A summary of the results in 1+1 dimensions and 2+1 dimensions is given in Sect. A.4.5.

A.4.1 Self-Affine Surface – Exponential Model

The exponential model for the autocorrelation function of a self-affine surface as discussed in Sect. 3.2 is given by

$$R(r) = \exp\left[-\left(\frac{r}{\xi}\right)^{2\alpha}\right]. \tag{A.55}$$

The PSD of a surface in 2+1 dimensions is related to the Hankel transform of the autocorrelation function,

$$P(k) = \frac{w^2}{2\pi}\int_0^\infty R(r)J_0(kr)r\mathrm{d}r = \frac{w^2}{2\pi}\mathcal{H}\left\{R(r)\right\}. \tag{A.56}$$

Using (A.6), the Hankel transform of the autocorrelation function is given by

$$\mathcal{H}\left\{\exp\left[-\left(\frac{r}{\xi}\right)^{2\alpha}\right]\right\} = \frac{1}{2\pi}\int_0^{2\pi}\int_0^\infty r\exp\left[-\left(\frac{r}{\xi}\right)^{2\alpha}\right]\exp\left[ikr\cos\theta\right]\mathrm{d}r\mathrm{d}\theta.$$

A Taylor expansion of the second exponential gives

$$\mathcal{H}\left\{\exp\left[-\left(\frac{r}{\xi}\right)^{2\alpha}\right]\right\} = \frac{1}{2\pi}\sum_{n=0}^{\infty}\int_0^{2\pi}\int_0^\infty r\exp\left[-\left(\frac{r}{\xi}\right)^{2\alpha}\right]\frac{(ikr\cos\theta)^n}{n!}\mathrm{d}r\mathrm{d}\theta.$$

Rearranging the integrals gives

$$\mathcal{H}\left\{\exp\left[-\left(\frac{r}{\xi}\right)^{2\alpha}\right]\right\} = \sum_{n=0}^{\infty}\int_0^{2\pi}\frac{(ik\cos\theta)^n}{2\pi n!}\mathrm{d}\theta\int_0^\infty r^{n+1}\exp\left[-\left(\frac{r}{\xi}\right)^{2\alpha}\right]\mathrm{d}r. \tag{A.57}$$

Carrying out the angular integration with (A.43), this expression becomes

$$\mathcal{H}\left\{\exp\left[-\left(\frac{r}{\xi}\right)^{2\alpha}\right]\right\} = \sum_{j=0}^{\infty}\frac{1}{(j!)^2}\left(-\frac{k^2}{4}\right)^j\int_0^\infty r^{2j+1}\exp\left[-\left(\frac{r}{\xi}\right)^{2\alpha}\right]\mathrm{d}r.$$

However, we can introduce the change of variable $s = (r/\xi)^{2\alpha}$ to write the Hankel transform as

$$\mathcal{H}\left\{\exp\left[-\left(\frac{r}{\xi}\right)^{2\alpha}\right]\right\} = \frac{\xi^2}{2\alpha}\sum_{j=0}^{\infty}\frac{1}{(j!)^2}\left(-\frac{k^2\xi^2}{4}\right)^j\int_0^\infty s^{(j+1)/\alpha-1}\mathrm{e}^{-s}\mathrm{d}s.$$

The integral can be written in terms of the gamma function as

$$\mathcal{H}\left\{\exp\left[-\left(\frac{r}{\xi}\right)^{2\alpha}\right]\right\} = \frac{\xi^2}{2\alpha}\sum_{j=0}^{\infty}\frac{\Gamma\left(\frac{j+1}{\alpha}\right)}{(j!)^2}\left(-\frac{k^2\xi^2}{4}\right)^j, \tag{A.58}$$

which leads to the PSD

$$P(k) = \frac{w^2\xi^2}{4\pi\alpha}\sum_{j=0}^{\infty}\frac{\Gamma\left(\frac{j+1}{\alpha}\right)}{(j!)^2}\left(-\frac{k^2\xi^2}{4}\right)^j. \tag{A.59}$$

For $\alpha = 1$, the sum can be explicitly computed, and the exponential model predicts the self-affine PSD in 2+1 dimensions to be

$$P(k) = \frac{w^2\xi^2}{4\pi}\exp\left[-\frac{k^2\xi^2}{4}\right]. \tag{A.60}$$

However, this form does not behave well as $\alpha \to 0$, and we investigate the K-correlation model in the next section that better deals with this behavior.

A.4.2 Self-Affine Surface – *K*-Correlation Model

The form of the autocorrelation function proposed by Palasantzas [118] for a self-affine surface is

$$R(r) = \frac{1}{a\Gamma(\alpha+1)}\left(\frac{r}{2\xi\sqrt{a}}\right)^{\alpha} K_{\alpha}\left(\frac{r}{\xi\sqrt{a}}\right), \tag{A.61}$$

where $K_\alpha(x)$ is the α-ordered modified Bessel function of the second kind introduced in Sect. A.1.3, and a is defined implicitly by

$$a = \frac{1}{2\alpha}\left[1 - \frac{1}{(1+ak_c^2\xi^2)^{\alpha}}\right], \tag{A.62}$$

where k_c is the upper cutoff frequency of the self-affine behavior for the surface, which would be inversely proportional to the lattice size. If we consider the continuum limit of no upper frequency cutoff, $k_c \to \infty$, we find that $a = (2\alpha)^{-1}$ for $\alpha \neq 0$, which gives

$$R(r) = \frac{\alpha}{2^{\alpha-1}\Gamma(\alpha+1)}\left(\frac{r}{\xi}\sqrt{2\alpha}\right)^{\alpha} K_{\alpha}\left(\frac{r}{\xi}\sqrt{2\alpha}\right). \tag{A.63}$$

The case where $\alpha = 0$ is obtained by taking the appropriate limit of the above expressions as $\alpha \to 0$, which gives

$$R(r) = \frac{1}{\tilde{a}}K_0\left(\frac{r}{\xi\sqrt{\tilde{a}}}\right) \quad \text{for } \alpha = 0, \tag{A.64}$$

where $\tilde{a}$ satisfies

$$\tilde{a} = \frac{1}{2}\ln\left(1+\tilde{a}k_c^2\xi^2\right). \tag{A.65}$$

Note that when $\alpha = 0$, taking the limit as $k_c \to \infty$ leads to a divergence, and the discrete nature of the surface profile must be taken into account by considering k_c when $\alpha = 0$. We are concerned only with the continuum limit of the autocorrelation function, and thus restrict the discussion to the interval $0 < \alpha \leq 1$. However, it is important to note that this model behaves correctly in the limit as $\alpha \to 0$ because $K_0(x) \sim -\ln x$ for small x, as opposed to the exponential correlation model which does not.

To find the PSD for $0 < \alpha \leq 1$, we need to find the Hankel transform of the autocorrelation function given in (A.63), which requires the evaluation of the integral

$$I = \int_0^\infty r^{\alpha+1} K_\alpha\left(\frac{r}{\xi}\sqrt{2\alpha}\right) J_0(kr)\mathrm{d}r. \tag{A.66}$$

To accomplish this, expand the Bessel function as (A.2),

$$\begin{aligned} I &= \int_0^\infty r^{\alpha+1} K_\alpha\left(\frac{r}{\xi}\sqrt{2\alpha}\right) \sum_{j=0}^{\infty} \frac{(-1)^j}{(j!)^2}\left(\frac{kr}{2}\right)^{2j} \mathrm{d}r \\ &= \sum_{j=0}^{\infty} \frac{(-1)^j}{(j!)^2}\left(\frac{k}{2}\right)^{2j} \int_0^\infty r^{\alpha+1+2j} K_\alpha\left(\frac{r}{\xi}\sqrt{2\alpha}\right) \mathrm{d}r. \end{aligned}$$

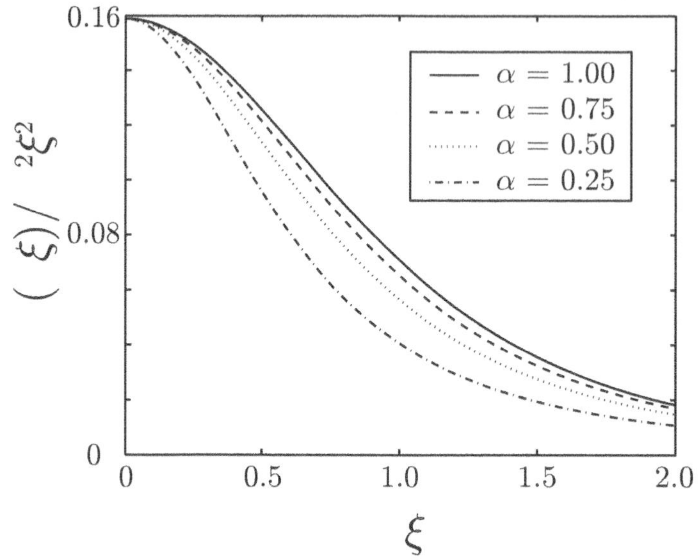

Fig. A.3. Sample self-affine PSD profiles from (A.68) for different values of α, including graphs with $\alpha = 1.00$, 0.75, 0.50, and 0.25.

Using the formula given by Watson [168],

$$\int_0^\infty K_\nu(x) x^{\mu-1} \mathrm{d}x = 2^{\mu-2} \Gamma\left(\frac{\mu-\nu}{2}\right) \Gamma\left(\frac{\mu+\nu}{2}\right),$$

the integral becomes

$$\begin{aligned} I &= \sum_{j=0}^{\infty} \frac{(-1)^j}{(j!)^2} \left(\frac{k}{2}\right)^{2j} \left[\left(\frac{\xi}{\sqrt{2\alpha}}\right)^{2j+\alpha+2} 2^{2j+\alpha} \Gamma(j+1) \Gamma(\alpha+j+1) \right] \\ &= \left(\frac{\xi}{\sqrt{2\alpha}}\right)^{\alpha+2} 2^\alpha \sum_{j=0}^{\infty} \frac{(-1)^j \Gamma(j+\alpha+1)}{j!} \left(\frac{k\xi}{\sqrt{2\alpha}}\right)^{2j} \\ &= \left(\frac{\xi}{\sqrt{2\alpha}}\right)^{\alpha+2} 2^\alpha \Gamma(\alpha+1) \sum_{j=0}^{\infty} \frac{\Gamma(j+\alpha+1)}{j! \Gamma(\alpha+1)} \left(-\frac{k^2\xi^2}{2\alpha}\right)^j . \end{aligned}$$

However, the series expansion for $(1+x^2)^{-\beta}$ is

$$\frac{1}{(1+x^2)^\beta} = \sum_{j=0}^{\infty} \frac{\Gamma(j+\beta)}{j!\Gamma(\beta)} \left(-x^2\right)^j ,$$

which means the integral can be written as

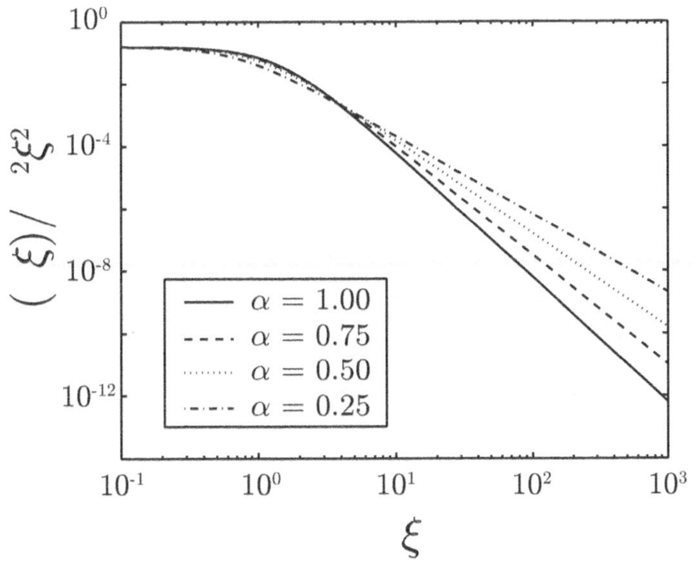

Fig. A.4. Sample self-affine PSD profiles on a log–log scale from (A.68) for different values of α, including graphs with $\alpha = 1.00$, 0.75, 0.50, and 0.25. The behavior of the PSD for large k is given by $P(k) \propto k^{-2-2\alpha}$.

$$I = \left(\frac{\xi}{\sqrt{2\alpha}}\right)^{\alpha+2} \frac{2^{\alpha}\Gamma(\alpha+1)}{\left(1+\frac{k^2\xi^2}{2\alpha}\right)^{1+\alpha}}.$$

Therefore, the Hankel transform of the autocorrelation function given in (A.63) is

$$\mathcal{H}\left\{\frac{\alpha}{2^{\alpha-1}\Gamma(\alpha+1)}\left(\frac{r}{\xi}\sqrt{2\alpha}\right)^{\alpha} K_{\alpha}\left(\frac{r}{\xi}\sqrt{2\alpha}\right)\right\} = \frac{\xi^2}{\left(1+\frac{k^2\xi^2}{2\alpha}\right)^{1+\alpha}}. \quad (A.67)$$

The PSD is $w^2/(2\pi)$ times the Hankel transform of the autocorrelation function from (2.22), which gives

$$P(k) = \frac{w^2\xi^2}{2\pi\left(1+\frac{k^2\xi^2}{2\alpha}\right)^{1+\alpha}}. \quad (A.68)$$

The full width at half maximum (FWHM) of this PSD is given by

$$\text{FWHM} = \frac{\sqrt{2\alpha(2^{1/(1+\alpha)}-1)}}{\xi}. \quad (A.69)$$

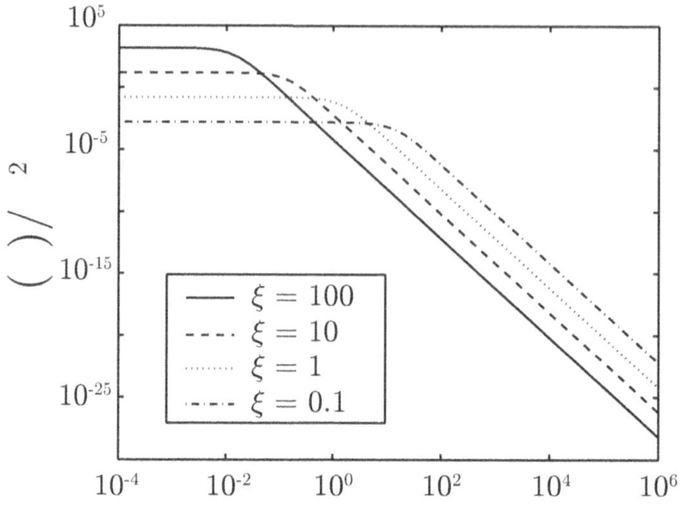

Fig. A.5. Sample self-affine PSD profiles on a log–log scale with different values for ξ from (A.68). Larger values of the correlation length ξ result in a smaller full width at half maximum (FWHM) as predicted by (A.69). All graphs have $\alpha = 1$.

For smaller values of the correlation length ξ, the PSD has a larger spread in frequency space, as seen in Fig. A.5. In addition, the behavior of the PSD for various values of α is shown in Fig. A.3, and on a log–log scale in Fig. A.4. The behavior of the PSD for large k is given by $P(k) \propto k^{-2-2\alpha}$, and thus the slope of the PSD on a log–log scale for large k will be equal to $-2 - 2\alpha$.

A.4.3 Mounded Surface – Exponential Model

The autocorrelation function for a mounded surface in 2+1 dimensions under the exponential model is

$$R(r) = \exp\left[-\left(\frac{r}{\xi}\right)^{2\alpha}\right] J_0\left(\frac{2\pi r}{\lambda}\right). \tag{A.70}$$

Because this autocorrelation function is the product of two functions, the PSD is proportional to the convolution of the two Hankel transforms from (A.53),

$$P(k) = \frac{w^2}{(2\pi)^2}\mathcal{H}\left\{\exp\left[-\left(\frac{r}{\xi}\right)^{2\alpha}\right]\right\} * \mathcal{H}\left\{J_0\left(\frac{2\pi r}{\lambda}\right)\right\}. \tag{A.71}$$

The Hankel transform of the Bessel function is given by (A.5),

$$\mathcal{H}\left\{J_0\left(\frac{2\pi r}{\lambda}\right)\right\} = \int_0^\infty r J_0\left(\frac{2\pi r}{\lambda}\right) J_0(kr)\mathrm{d}r = \frac{1}{k}\delta\left(k - \frac{2\pi}{\lambda}\right). \tag{A.72}$$

This relation is why the Bessel function is introduced in the autocorrelation function in (A.70) because the convolution integral in (A.71) is much simpler if one of the transforms is a delta function.

The Hankel transform of the exponential is more difficult, and we examine the case where $\alpha = 1$. The Hankel transform of the exponential when $\alpha = 1$ was found in Sect. A.4.1, and utilized in (A.60). The PSD is then given by the convolution of the two Hankel transforms. We must take care to take the convolution in Cartesian coordinates and then convert to polar to carry out the integral.

$$\begin{aligned} P(k) = \frac{w^2\xi^2}{8\pi^2} \int\int \exp\left[-\frac{\xi^2}{4}\left((k_x - x')^2 + (k_y - y')^2\right)\right] \\ \times \frac{\delta\left(\sqrt{(x')^2 + (y')^2} - \frac{2\pi}{\lambda}\right)}{\sqrt{(x')^2 + (y')^2}} \mathrm{d}x'\mathrm{d}y'. \end{aligned}$$

Changing variables from Cartesian coordinates (x', y') to polar coordinates (r, θ), $\mathrm{d}x'\mathrm{d}y' = r\mathrm{d}r\mathrm{d}\theta$, gives

$$\begin{aligned} P(k) = & \frac{w^2\xi^2}{8\pi^2} \int_0^{2\pi} \int_0^\infty \exp\left[-\frac{\xi^2}{4}\left((k_x - r\cos\theta)^2 + (k_y - r\sin\theta)^2\right)\right] \\ & \times\delta\left(r - \frac{2\pi}{\lambda}\right) \mathrm{d}r\mathrm{d}\theta. \end{aligned}$$

Integrating over the delta function,

$$P(k) = \frac{w^2\xi^2}{8\pi^2} \int_0^{2\pi} \exp\left[-\frac{\xi^2}{4}\left(\left(k_x - \frac{2\pi}{\lambda}\cos\theta\right)^2 + \left(k_y - \frac{2\pi}{\lambda}\sin\theta\right)^2\right)\right] \mathrm{d}\theta.$$

Expanding the argument of the exponential gives

$$P(k) = \frac{w^2\xi^2}{8\pi^2} \int_0^{2\pi} \exp\left[-\frac{\xi^2}{4}\left(k^2 + \frac{4\pi^2}{\lambda^2} - \frac{4\pi k_x}{\lambda}\cos\theta - \frac{4\pi k_y}{\lambda}\sin\theta\right)\right] \mathrm{d}\theta.$$

This can be simplified with the relation

$$a\cos\theta + b\sin\theta = \sqrt{a^2 + b^2}\cos\left(\theta - \tan^{-1}\frac{b}{a}\right),$$

which follows from the expansion of $\cos(x + y) = \cos x\cos y - \sin x\sin y$. Because the integral over θ is over a complete period of $\cos\theta$, the phase shift of $\tan^{-1} b/a$ does not affect the integral,

$$P(k) = \frac{w^2\xi^2}{8\pi^2}\int_0^{2\pi}\exp\left[-\frac{\xi^2}{4}\left(k^2+\frac{4\pi^2}{\lambda^2}-\frac{4\pi k}{\lambda}\cos\theta\right)\right]\mathrm{d}\theta$$
$$= \frac{w^2\xi^2}{8\pi^2}\exp\left[-\frac{\xi^2}{4}\left(k^2+\frac{4\pi^2}{\lambda^2}\right)\right]\int_0^{2\pi}\exp\left[\frac{\pi k\xi^2}{\lambda}\cos\theta\right]\mathrm{d}\theta.$$

From (A.11), this can be written as

$$P(k) = \frac{w^2\xi^2}{4\pi}\exp\left[-\frac{\left(4\pi^2+k^2\lambda^2\right)\xi^2}{4\lambda^2}\right] I_0\left(\frac{\pi k\xi^2}{\lambda}\right). \tag{A.73}$$

This functional form assumes that $\alpha = 1$. For $\alpha \neq 1$, the Hankel transform of the exponential in the autocorrelation function can be expressed as a series from (A.58),

$$\mathcal{H}\left\{\exp\left[-\left(\frac{r}{\xi}\right)^{2\alpha}\right]\right\} = \frac{\xi^2}{2\alpha}\sum_{j=0}^{\infty}\frac{\Gamma\left(\frac{j+1}{\alpha}\right)}{(j!)^2}\left(-\frac{k^2\xi^2}{4}\right)^j. \tag{A.74}$$

There is clearly a problem when $\alpha = 0$, and this is addressed in the next section. Carrying out similar steps to the calculation with $\alpha = 1$, the PSD for general α can be expressed as a series,

$$P(k) = \frac{w^2\xi^2}{8\pi^2\alpha}\sum_{j=0}^{\infty}\frac{\Gamma\left(\frac{j+1}{\alpha}\right)}{(j!)^2}\left[-\frac{\pi k\xi^2}{a\lambda}\right]^j\int_0^{2\pi}[1-a\cos\theta]^j\mathrm{d}\theta, \tag{A.75}$$

where $a = 4\pi\lambda k/\left(k^2\lambda^2+4\pi^2\right)$. This formula can be used for a numerical computation of the PSD for various values of α. We forgo further analysis of this form of the PSD because of the behavior for small α and introduce another autocorrelation function that is valid as $\alpha \to 0$.

A.4.4 Mounded Surface – *K*-Correlation Model

Consider the autocorrelation function

$$R(r) = \frac{\alpha}{2^{\alpha-1}\Gamma(\alpha+1)}\left(\frac{r}{\xi}\sqrt{2\alpha}\right)^{\alpha}K_\alpha\left(\frac{r}{\xi}\sqrt{2\alpha}\right)J_0\left(\frac{2\pi r}{\lambda}\right). \tag{A.76}$$

This is a combination of the autocorrelation function from the K-correlation model proposed for self-affine surfaces in Sect. A.4.2, along with the Bessel function introduced to model mounded surfaces. The advantage of this autocorrelation function is that we already know the Hankel transforms for both parts,

$$\mathcal{H}\left\{\frac{\alpha}{2^{\alpha-1}\Gamma(\alpha+1)}\left(\frac{r}{\xi}\sqrt{2\alpha}\right)^{\alpha}K_\alpha\left(\frac{r}{\xi}\sqrt{2\alpha}\right)\right\} = \frac{\xi^2}{\left(1+\frac{k^2\xi^2}{2\alpha}\right)^{1+\alpha}}, \tag{A.77}$$

$$\mathcal{H}\left\{J_0\left(\frac{2\pi r}{\lambda}\right)\right\} = \frac{1}{k}\delta\left(k-\frac{2\pi}{\lambda}\right). \tag{A.78}$$

There are two reasons to consider this model for the autocorrelation function over the exponential model introduced in Sect. A.4.3. First, the exponential model for a mounded PSD diverges as $\alpha \to 0$, and the K-correlation model will remedy this behavior. Also, the functional form of the PSD for $\alpha = 1$ from the exponential model involves transcendental functions as given in (A.73), and is hard to analyze. We show that this new autocorrelation function gives a PSD that is a rational function for $\alpha = 1$. This allows us to analytically solve for the peak position k_m and see that it is equal to $2\pi\lambda^{-1}$ under certain conditions.

The Hankel transforms are already known, thus we can write the PSD as the convolution

$$P(k) = \frac{w^2}{4\pi^2} \int\int \frac{\xi^2}{\left[1 + \frac{\xi^2}{2\alpha}\left((k_x - x')^2 + (k_y - y')^2\right)\right]^{1+\alpha}} \times \frac{\delta\left(\sqrt{(x')^2 + (y')^2} - \frac{2\pi}{\lambda}\right)}{\sqrt{(x')^2 + (y')^2}} \mathrm{d}x'\mathrm{d}y'.$$

Transforming to polar coordinates gives

$$P(k) = \frac{w^2\xi^2}{4\pi^2} \int_0^{2\pi} \frac{\mathrm{d}\theta}{\left[1 + \frac{\xi^2}{2\alpha}\left(k^2 + \frac{4\pi^2}{\lambda^2} - \frac{4\pi k}{\lambda}\cos\theta\right)\right]^{1+\alpha}}.$$

It follows that the PSD is given by

$$P(k) = \frac{w^2\xi^2}{4\pi^2} \frac{1}{\left[1 + \frac{k^2\xi^2}{2\alpha} + \frac{2\pi^2\xi^2}{\alpha\lambda^2}\right]^{1+\alpha}} \int_0^{2\pi} \frac{\mathrm{d}\theta}{[1 - a\cos\theta]^{1+\alpha}}, \tag{A.79}$$

where

$$a = \frac{\frac{2\pi\xi^2 k}{\alpha\lambda}}{1 + \frac{k^2\xi^2}{2\alpha} + \frac{2\pi^2\xi^2}{\alpha\lambda^2}}.$$

This mounded PSD is pictured in Fig. A.6 for various values of α, and on a log–log scale in Fig. A.7. We first examine this equation with $\alpha = 1$, then discuss the expression for general α. For $\alpha = 1$, we need to use the result

$$\int_0^{2\pi} \frac{\mathrm{d}\theta}{(1 - a\cos\theta)^2} = \frac{2\pi}{(1 - a^2)^{3/2}}, \tag{A.80}$$

which can be found using complex variable techniques. The PSD becomes

$$P(k) = \frac{w^2\xi^2}{2\pi\left[1 + \frac{k^2\xi^2}{2} + \frac{2\pi^2\xi^2}{\lambda^2}\right]^2} \frac{1}{\left[1 - \left(\frac{\frac{2\pi\xi^2 k}{\lambda}}{1 + \frac{k^2\xi^2}{2} + \frac{2\pi^2\xi^2}{\lambda^2}}\right)^2\right]^{3/2}} \tag{A.81}$$

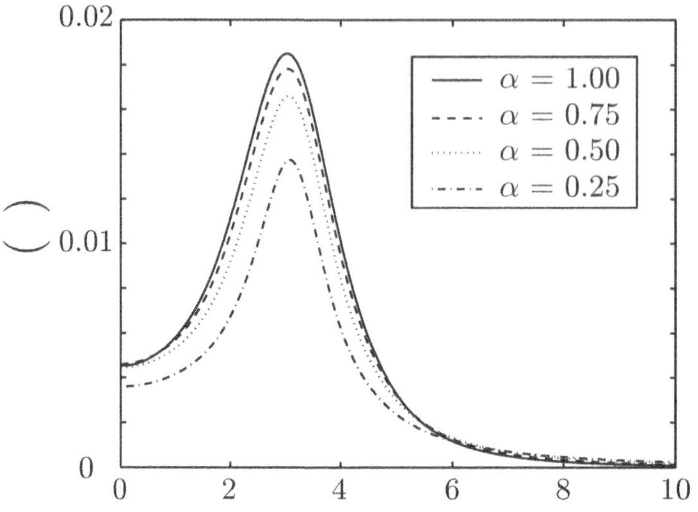

Fig. A.6. Sample mounded PSD profiles for different values of α from (A.79), including graphs with $\alpha = 1.00$, 0.75, 0.50, and 0.25. All graphs have $\lambda = 2$, $\xi = 1$, and $w = 1$. Note that the peak position lies at $k_m \approx \pi$, which corresponds to $2\pi/\lambda$.

$$= \frac{w^2\xi^2}{2\pi} \frac{1 + \frac{2\pi^2\xi^2}{\lambda^2} + \frac{\xi^2}{2}k^2}{\left[\left(1 + \frac{2\pi^2\xi^2}{\lambda^2} + \frac{\xi^2}{2}k^2\right)^2 - \left(\frac{2\pi\xi^2}{\lambda}k\right)^2\right]^{3/2}}. \tag{A.82}$$

This is the analytical form of the PSD for $\alpha = 1$, and it is expressed as a rational fraction. Now that we have an analytical PSD at our disposal, let us investigate the relation $k_m = 2\pi\lambda^{-1}$. To show this, we must set the derivative of the PSD with respect to k equal to zero and solve for k_m. Because the PSD is a rational function, the derivative will also be a rational function, and k_m can be obtained by setting the numerator of the derivative equal to zero. After some tedious algebra, one obtains

$$\lambda^4\xi^4k_m^4 + 4\lambda^2\xi^2(\pi^2\xi^2 + \lambda^2)k_m^2 - 4(2\pi^2\xi^2 + \lambda^2)(4\pi^2\xi^2 - \lambda^2) = 0.$$

This is a quadratic equation in k_m^2, with solution

$$k_m^2 = \frac{2\pi\xi\sqrt{4\lambda^2 + 9\pi^2\xi^2} - 2\pi^2\xi^2 - 2\lambda^2}{\lambda^2\xi^2}.$$

We are only concerned with positive wavenumbers, so k_m is given by

$$k_m = \frac{\sqrt{2\pi\xi\sqrt{4\lambda^2 + 9\pi^2\xi^2} - 2\pi^2\xi^2 - 2\lambda^2}}{\lambda\xi}.$$

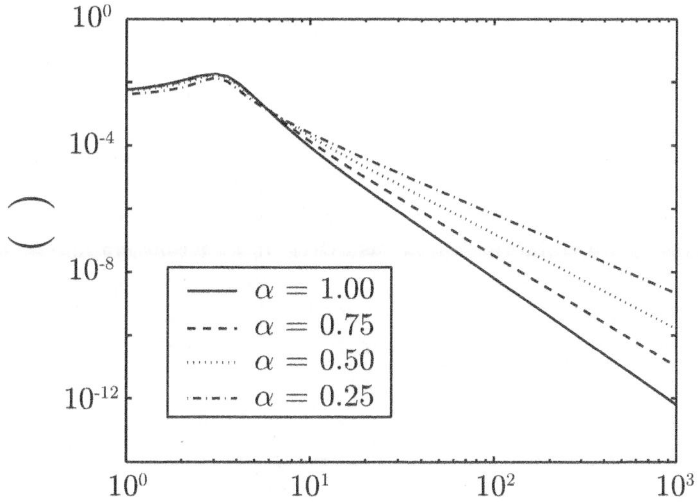

Fig. A.7. Sample mounded PSD profiles on a log–log scale for different values of α from (A.79), including graphs with $\alpha = 1.00$, 0.75, 0.50, and 0.25. All graphs have $\lambda = 2$, $\xi = 1$, and $w = 1$. The behavior of the PSD for large k is given by $P(k) \propto k^{-2-2\alpha}$.

This is certainly more complicated than the simple relation $k_m = 2\pi\lambda^{-1}$. However, with a rearrangement we obtain

$$k_m = \frac{\sqrt{6\pi^2\sqrt{1+\frac{4\lambda^2}{9\pi^2\xi^2}} - 2\pi^2 - \frac{2\lambda^2}{\xi^2}}}{\lambda}.$$

We can Taylor expand the innermost square root to obtain

$$k_m = \frac{\sqrt{4\pi^2 - \frac{2\lambda^2}{3\xi^2} + O\left(\frac{\lambda^4}{\xi^4}\right)}}{\lambda} = \frac{2\pi\sqrt{1 - \frac{\lambda^2}{6\pi^2\xi^2} + O\left(\frac{\lambda^4}{\xi^4}\right)}}{\lambda}.$$

Again Taylor expanding gives

$$k_m = \frac{2\pi}{\lambda}\left(1 - \frac{\lambda^2}{12\pi^2\xi^2} + O\left(\frac{\lambda^4}{\xi^4}\right)\right). \tag{A.83}$$

This gives us a condition when $k_m \approx 2\pi\lambda^{-1}$,

$$\frac{\lambda}{\xi} \ll 2\pi\sqrt{3} \approx 10.88. \tag{A.84}$$

Ordinarily, the wavelength is defined to be $\lambda \equiv 2\pi k_m^{-1}$, and the last result simply shows that (A.76) is consistent with this definition if the wavelength is not an order of magnitude larger than the correlation length. A different model for the autocorrelation function could give a better agreement with the relation $k_m = 2\pi\lambda^{-1}$, so there is room for improvement of this proposed model.

The full width at half maximum (FWHM) of this PSD is more complicated to determine, but in the regime where $\pi\xi\sqrt{2} \ll \lambda$, the PSD is a function of $k\xi$ only, which implies that FWHM $\propto \xi^{-1}$. Previous work has shown this behavior to be true through a numerical analysis [187]. It is a reasonable conclusion because the FWHM for a self-affine PSD is also inversely proportional to the lateral correlation length, as was shown in (A.69).

For a general value of α, expanding the integrand in (A.79) and using (A.43),

$$P(k) = \frac{w^2\xi^2}{2\pi\Gamma(\alpha+1)\left[1 + \frac{k^2\xi^2}{2\alpha} + \frac{2\pi^2\xi^2}{\alpha\lambda^2}\right]^{1+\alpha}} \sum_{j=0}^{\infty} \frac{\Gamma(2j+\alpha+1)}{(j!)^2}\left(\frac{a^2}{4}\right)^j, \tag{A.85}$$

where

$$a = \frac{\frac{2\pi\xi^2 k}{\alpha\lambda}}{1 + \frac{k^2\xi^2}{2\alpha} + \frac{2\pi^2\xi^2}{\alpha\lambda^2}}.$$

This form does not diverge as $\alpha \to 0$ as can be seen from an appropriate rearrangement of the PSD,

$$P(k) = \frac{w^2\xi^2(2\alpha)^{1+\alpha}}{2\pi\Gamma(\alpha+1)\left[2\alpha + k^2\xi^2 + \frac{4\pi^2\xi^2}{\lambda^2}\right]^{1+\alpha}} \sum_{j=0}^{\infty} \frac{\Gamma(2j+\alpha+1)}{(j!)^2}\left(\frac{a^2}{4}\right)^j, \tag{A.86}$$

where

$$a = \frac{4\pi\xi^2\lambda k}{2\alpha\lambda^2 + k^2\xi^2\lambda^2 + 4\pi^2\xi^2}.$$

However, recall from Sect. A.4.2 that the behavior of the PSD when $\alpha = 0$ must be treated separately in terms of the cutoff frequency k_c, which is not discussed here. A plot of this form of the PSD for various values of α is included in Fig. A.6. Changing the value of α does not significantly affect the peak position or the width of the peak, although smaller values of α yield a less pronounced peak intensity.

In the limit of large wavenumbers, $k \gg 2\pi/\lambda$, $a \approx 4\pi/(k\lambda)$, and the sum in (A.85) can be approximated by the first term. This gives

$$P(k \gg 2\pi/\lambda) \approx \frac{w^2\xi^2}{2\pi\left(1 + \frac{k^2\xi^2}{2\alpha}\right)^{1+\alpha}}, \tag{A.87}$$

which is the form of a self-affine PSD from (A.68). This behavior is seen in Fig. A.7, which is graphed on a log–log scale. On a logarithmic scale, the slope of the graph for large k is $-2-2\alpha$. Therefore, at wavenumbers much larger than $2\pi/\lambda$, or equivalently at length scales much smaller than λ, the surface appears self-affine. Similarly, in the limit as $\lambda \to \infty$, $a \to 0$; and the PSD approaches the form of a self-affine PSD.

A.4.5 Summary

We present here a summary of the results for various forms of the power spectral density function in 1+1 and 2+1 dimensions. The mathematics of computing the PSD is similar in 1+1 dimensions to that presented in the previous sections for 2+1 dimensions. In 1+1 dimensions, the PSD can be expressed as

$$\begin{aligned} P(k) &= \frac{w^2}{2\pi}\int_{-\infty}^{\infty} R(r)\mathrm{e}^{-\mathrm{i}kr}\mathrm{d}r \\ &= \frac{w^2}{\pi}\int_{0}^{\infty} R(r)\cos kr\mathrm{d}r, \end{aligned}$$

where the integration over the imaginary part of the exponential drops out because the autocorrelation function $R(r)$ is even. By expanding $\cos kr$ in a Taylor series and integrating, the results for the PSD in 1+1 dimensions can be obtained. Because the mathematics is very similar to the derivations already presented, we simply summarize the results here. The only difference in the autocorrelation functions in 1+1 dimensions and 2+1 dimensions is the presence of a cos term in the mounded models instead of a Bessel function, which allows for the analytical expressions that follow.

1+1 Dimensions

- Self-affine exponential model:

$$R(r) = \exp\left[-\left(\frac{r}{\xi}\right)^{2\alpha}\right], \tag{A.88}$$

$$P(k) = \frac{w^2\xi}{2\pi\alpha}\sum_{j=0}^{\infty}\frac{\Gamma\left(\frac{j+1/2}{\alpha}\right)}{(2j)!}\left(-k^2\xi^2\right)^j. \tag{A.89}$$

- Self-affine exponential model ($\alpha = 1$):

$$R(r) = \exp\left[-\left(\frac{r}{\xi}\right)^{2}\right], \tag{A.90}$$

$$P(k) = \frac{w^2\xi}{2\sqrt{\pi}}\exp\left[-\frac{k^2\xi^2}{4}\right]. \tag{A.91}$$

- Self-affine K-correlation model:

$$R(r) = \frac{\alpha}{2^{\alpha-1}\Gamma(\alpha+1)} \left(\frac{r}{\xi}\sqrt{2\alpha}\right)^{\alpha} K_{\alpha}\left(\frac{r}{\xi}\sqrt{2\alpha}\right), \tag{A.92}$$

$$P(k) = \frac{w^2 \xi}{\sqrt{2\pi}} \frac{\sqrt{\alpha}\Gamma(\alpha+1/2)}{\Gamma(\alpha+1)} \left(1 + \frac{k^2\xi^2}{2\alpha}\right)^{-\alpha-1/2} . \tag{A.93}$$

- Mounded exponential model:

$$R(r) = \exp\left[-\left(\frac{r}{\xi}\right)^{2\alpha}\right] \cos\left(\frac{2\pi r}{\lambda}\right), \tag{A.94}$$

$$P(k) = \frac{w^2\xi}{4\pi\alpha} \sum_{j=0}^{\infty} \frac{\Gamma\left(\frac{j+1/2}{\alpha}\right)}{(2j)!} \left(-\xi^2\right)^j \left[\left(k - \frac{2\pi}{\lambda}\right)^{2j} + \left(k + \frac{2\pi}{\lambda}\right)^{2j}\right]. \tag{A.95}$$

- Mounded exponential model ($\alpha = 1$):

$$R(r) = \exp\left[-\left(\frac{r}{\xi}\right)^{2}\right] \cos\left(\frac{2\pi r}{\lambda}\right), \tag{A.96}$$

$$P(k) = \frac{w^2\xi}{2\sqrt{\pi}} \exp\left[-\frac{\left(4\pi^2 + k^2\lambda^2\right)\xi^2}{4\lambda^2}\right] \cosh\left(\frac{\pi k \xi^2}{\lambda}\right). \tag{A.97}$$

- Mounded K-correlation model:

$$R(r) = \frac{\alpha}{2^{\alpha-1}\Gamma(\alpha+1)} \left(\frac{r}{\xi}\sqrt{2\alpha}\right)^{\alpha} K_{\alpha}\left(\frac{r}{\xi}\sqrt{2\alpha}\right) \cos\left(\frac{2\pi r}{\lambda}\right), \tag{A.98}$$

$$P(k) = \frac{w^2\xi}{2\sqrt{2\pi}} \frac{\sqrt{\alpha}\Gamma(\alpha+1/2)}{\Gamma(\alpha+1)} \left[\left(1 + \frac{\left(k - \frac{2\pi}{\lambda}\right)^2 \xi^2}{2\alpha}\right)^{-\alpha-1/2} + \left(1 + \frac{\left(k + \frac{2\pi}{\lambda}\right)^2 \xi^2}{2\alpha}\right)^{-\alpha-1/2}\right]. \tag{A.99}$$

2+1 Dimensions

- Self-affine exponential model:

$$R(r) = \exp\left[-\left(\frac{r}{\xi}\right)^{2\alpha}\right], \tag{A.100}$$

$$P(k) = \frac{w^2\xi^2}{4\pi\alpha}\sum_{j=0}^{\infty}\frac{\Gamma\left(\frac{j+1}{\alpha}\right)}{(j!)^2}\left(-\frac{k^2\xi^2}{4}\right)^j. \tag{A.101}$$

- Self-affine exponential model ($\alpha = 1$):

$$R(r) = \exp\left[-\left(\frac{r}{\xi}\right)^2\right], \tag{A.102}$$

$$P(k) = \frac{w^2\xi^2}{4\pi}\exp\left[-\frac{k^2\xi^2}{4}\right]. \tag{A.103}$$

- Self-affine K-correlation model:

$$R(r) = \frac{\alpha}{2^{\alpha-1}\Gamma(\alpha+1)}\left(\frac{r}{\xi}\sqrt{2\alpha}\right)^{\alpha} K_\alpha\left(\frac{r}{\xi}\sqrt{2\alpha}\right), \tag{A.104}$$

$$P(k) = \frac{w^2\xi^2}{2\pi\left(1+\frac{k^2\xi^2}{2\alpha}\right)^{1+\alpha}}. \tag{A.105}$$

- Mounded exponential model:

$$R(r) = \exp\left[-\left(\frac{r}{\xi}\right)^{2\alpha}\right] J_0\left(\frac{2\pi r}{\lambda}\right), \tag{A.106}$$

$$P(k) = \frac{w^2\xi^2}{8\pi^2\alpha}\sum_{j=0}^{\infty}\frac{\Gamma\left(\frac{j+1}{\alpha}\right)}{(j!)^2}\left[-\frac{\pi k\xi^2}{a\lambda}\right]^j\int_0^{2\pi}[1-a\cos\theta]^j\mathrm{d}\theta,$$

$$\text{where } a = \frac{4\pi\lambda k}{k^2\lambda^2+4\pi^2}. \tag{A.107}$$

- Mounded exponential model ($\alpha = 1$):

$$R(r) = \exp\left[-\left(\frac{r}{\xi}\right)^2\right] J_0\left(\frac{2\pi r}{\lambda}\right), \tag{A.108}$$

$$P(k) = \frac{w^2\xi^2}{4\pi}\exp\left[-\frac{\left(4\pi^2+k^2\lambda^2\right)\xi^2}{4\lambda^2}\right] I_0\left(\frac{\pi k\xi^2}{\lambda}\right). \tag{A.109}$$

- Mounded K-correlation model:

$$R(r) = \frac{\alpha}{2^{\alpha-1}\Gamma(\alpha+1)}\left(\frac{r}{\xi}\sqrt{2\alpha}\right)^{\alpha} K_\alpha\left(\frac{r}{\xi}\sqrt{2\alpha}\right) J_0\left(\frac{2\pi r}{\lambda}\right), \tag{A.110}$$

$$P(k) = \frac{w^2\xi^2}{2\pi\Gamma(\alpha+1)\left[1+\frac{k^2\xi^2}{2\alpha}+\frac{2\pi^2\xi^2}{\alpha\lambda^2}\right]^{1+\alpha}} \sum_{j=0}^{\infty} \frac{\Gamma(2j+\alpha+1)}{(j!)^2}\left(\frac{a^2}{4}\right)^j,$$

$$\text{where } a = \frac{4\pi\xi^2\lambda k}{2\alpha\lambda^2+k^2\xi^2\lambda^2+4\pi^2\xi^2}. \tag{A.111}$$

- Mounded K-correlation model ($\alpha = 1$):

$$R(r) = \frac{r\sqrt{2}}{\xi}K_1\left(\frac{r\sqrt{2}}{\xi}\right) J_0\left(\frac{2\pi r}{\lambda}\right), \tag{A.112}$$

$$P(k) = \frac{w^2\xi^2}{2\pi}\frac{1+\frac{2\pi^2\xi^2}{\lambda^2}+\frac{\xi^2}{2}k^2}{\left[\left(1+\frac{2\pi^2\xi^2}{\lambda^2}+\frac{\xi^2}{2}k^2\right)^2-\left(\frac{2\pi\xi^2}{\lambda}k\right)^2\right]^{3/2}}. \tag{A.113}$$

B

Euler's Method Implementation

The following is a C++ implementation of Euler's method to compute the solution to the partial differential equation

$$\frac{\partial h(x,t)}{\partial t} = \nabla^2 h(x,t) + \eta(x,t), \tag{B.1}$$

in one spatial dimension with cyclic boundary conditions and Gaussian distributed noise. A representative graph of the surface profile is included in Fig. B.1, and the evolution of the interface width is pictured in Fig. B.2.

```
#include <iostream>
#include <fstream>
#include <cmath>

using namespace std;

float* h;   //Stores h(x) at the current time
float* h_old;   //Stores h(x) at the previous time iteration
float delta_x = 1;
float delta_t = 0.001;
int N = 5000;  //Total length of domain = N*delta_x
int T = 10000;  //Total length of time = T*delta_t
float D = 1;  //Variance of the noise

float mean = 0;
float roughness = 0;

float laplacian(int j) //Computes the Laplacian of h(x)
{
      if(j == 0) //Cyclic boundary conditions
       return (h[j+1]-2*h[j]+h[N-1])/((delta_x)*(delta_x));
      else if(j == N-1) //Cyclic boundary conditions
```

```
        return (h[0]-2*h[j]+h[j-1])/((delta_x)*(delta_x));
       else
        return (h[j+1]-2*h[j]+h[j-1])/((delta_x)*(delta_x));
}

float noise()
{
       //Box-Muller transform to generate Gaussian random
       //variables with mean zero and variance one
       float x1, x2, k, y;

          do {
                  x1 = 2 * rand()/((float)RAND_MAX) - 1;
                  x2 = 2 * rand()/((float)RAND_MAX) - 1;
                  k = x1*x1 + x2*x2;
          } while ( k >= 1.0 );

          k = sqrt((-2.0*log(k))/k);
          y = x1 * k;

       return y;
}

int main()
{
   h = new float[N];
   h_old = new float[N];
   //Seed the random number generator
   srand((unsigned)time( NULL ));

   ofstream stats;
   stats.open("stats.txt");

   for(int i = 0; i < N; i++)
   {
        h[i] = 0;
   }

   for(int i = 0; i < N; i++)
   {
        h_old[i] = h[i];
   }

   //Iterate through time steps
   for(int k = 0; k < T; k++)
```

```
{

    //Iterate through position steps
    for(int j = 0; j < N; j++)
    {
    //Euler's Method
    h[j] = h_old[j]+
      delta_t*(laplacian(j)+sqrt( 2*D/
      ((delta_t)*(delta_x)*(delta_x)) )*noise());

    }

    //Iterate to next time step
    for(int i = 0; i < N; i++)
    {
           h_old[i] = h[i];
    }

    mean = 0;

    for(int i = 0; i < N; i++)
    {
           mean += h[i];
    }

    mean = mean / N;

    roughness = 0;

    for(int i = 0; i < N; i++)
    {
           roughness += (h[i]-mean)*(h[i]-mean);
    }

    roughness = sqrt(roughness / N);

    stats << k << " " << mean << " "
          << roughness << endl;

}

ofstream output;       //Output the results
output.open("result.txt");
 for(int i = 0; i < N; i++)
{
```

```
        output << i << " " << h[i] << endl;
    }

    return 0;

}
```

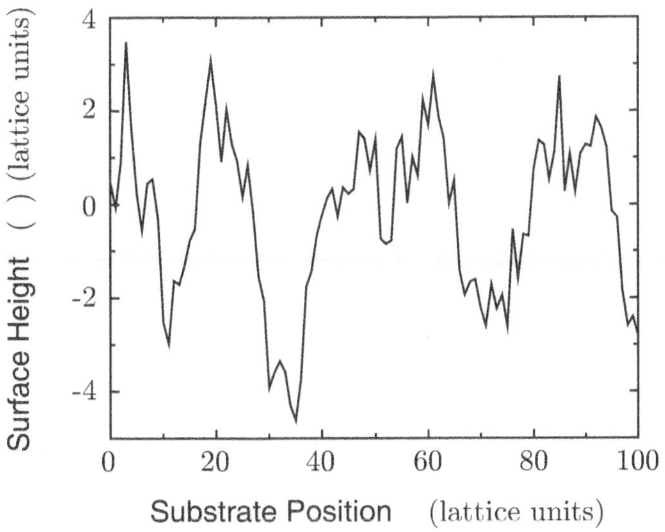

Fig. B.1. An example surface profile from the included code to model the Edwards–Wilkinson equation in one dimension. The position step size is $\Delta x = 1$, and the Gaussian noise has a variance $D = 1$.

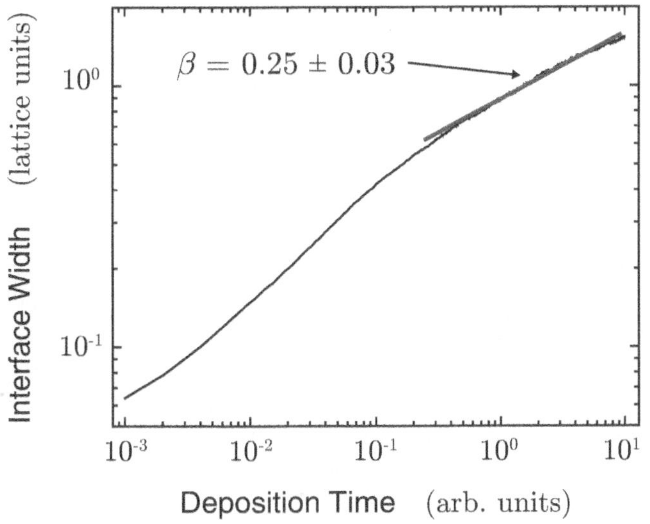

Fig. B.2. Interface width evolution of the Edwards–Wilkinson equation in one dimension calculated from the included code. The interface width evolves with an exponent $\beta = 0.25 \pm 0.03$, consistent with the discussion in Sect. 5.1.2 for $d = 1$.

C

Small World Model Implementation

The following is a C++ implementation of the small world model discussed in Chap. 6. This code simulates a small world network model with negative (shadowing) links between heights separated by a predetermined distance, $\lambda_0 = 10$ lattice units. This code can be modified to simulate surface growth under reemission and geometrical shadowing links as discussed in Chap. 6.

```
#include <iostream>
#include <cmath>
#include <fstream>
#include <time.h>
#include <string>

using namespace std;

const int N = 128;

float * Surface [N];
float * SurfaceNew [N];
float * links[N];

string filenm;
float D,deltat;
float deltax = 1;

float h(int x,int y)//Surface height
{
   while(x<0){x=x+N;}
   while(x>N-1){x=x-N;}

   while(y<0){y=y+N;}
   while(y>N-1){y=y-N;}
```

```
   return Surface[x][y];
}

float del2xy(int x, int y) //Laplacian
{
   return ( (h(x+1,y)+h(x-1,y)+
            h(x,y+1)+h(x,y-1)-
             (float)4*h(x,y) )/(deltax)*(deltax) );
}


float sw(int x, int y)
//Small world links for surface point at (x,y)
{
   float growth = 0;

    int shad_len = 10;
    //Shadowing length, or lambda_0
    //Heights are linked if separated by
    //a distance less than shad_len

    for(int k=-shad_len; k<=shad_len; k++)
    {
      for(int l=-shad_len; l<=shad_len; l++)
      {
           if( sqrt(k*k+l*l)<= shad_len)
            {
              growth +=  -(h(x+k,y+l)-h(x,y));
              //Link strength is negative
            }
      }
   }

   return growth;
}
float noise()
{
      //Box-Muller transform to generate Gaussian random
      //variables with mean zero and variance one
      float x1, x2, k, y;

         do {
                x1 = 2 * rand()/((float)RAND_MAX) - 1;
                x2 = 2 * rand()/((float)RAND_MAX) - 1;
```

```
                k = x1*x1 + x2*x2;
        } while ( k >= 1.0 );

        k = sqrt((-2.0*log(k))/k);
        y = x1 * k;

      return y;
}
void set_links() //Normalize link strength
{
     float sum = 0;
     float num = 0;

     for(int i = 0; i < N; i++)
     {
         for(int j = 0; j < N; j++)
         {
             num = sw(i,j);
             links[i][j]=num;
             //Find sum of link strengths
             sum += abs(num);
         }
     }

     if(sum == 0) return;

     //Normalize link strength
     for(int i = 0; i < N; i++)
     {
        for(int j = 0; j < N; j++)
        {
           links[i][j]=(links[i][j]/abs(sum))*N*N*25;
           //Choose strength of links to compete with
           //strength of random noise
        }
     }

     return;
}

float rms(float mean) //Surface roughness
{
   float roughness = 0;
   for(int i = 0; i < N; i++)
   {
```

```
        for(int j = 0; j < N; j++)
        {
            roughness += (h(i,j)-mean)*(h(i,j)-mean);
        }
    }
    return sqrt((double)roughness / (double)(N*N));
}
int main()
{
    //Create Lattice
    filenm = "s1";
    deltat=0.025;
    D = 100;
    srand((unsigned)time(NULL));
    ofstream output2,output3;

    output2.open(string(filenm+string("_")
    +string("rms.txt")).c_str());

    output3.open("mean.txt");

    for (int i = 0;i < N;i++)
    {
        Surface [i] = new float [N];
        for(int j =0; j < N; j++)
        {
            Surface[i][j] = sqrt((float)(2*D)*
                    (deltat)/( (deltax)*(deltax)) )*noise();
        }
    }

    for (int i = 0;i < N;i++)
    {
        SurfaceNew [i] = new float [N];
        if(SurfaceNew[i]==0)
        {
            cerr << "Memory Error";
        }
    }

    for (int i = 0;i < N;i++)
    {
        links [i] = new float [N];
        if(links[i]==0)
        {
```

```
      cerr << "Memory Error";
   }
}

for(int t=1; t <= 500;t++)
{
   cout << "t = " << t <<endl;

   set_links();  //Normalize link strength
   for(int i = 0; i<N; i++)
   {
    for(int j = 0; j < N; j++)
    {
      SurfaceNew[i][j] = Surface[i][j]
      + deltat*(del2xy(i,j) + links[i][j]
      + sqrt( (float)(2*D)
        /((deltat)*(deltax)*(deltax)) )*noise());
    }
   }

   for(int i=0; i<N; i++)
   {
      for(int j = 0; j < N; j++)
      {
         Surface[i][j] = SurfaceNew[i][j];
      }
   }

   float mean = 0;
   for(int i=0; i<N; i++)
   {
      for(int j=0; j<N; j++)
      {
        mean += h(i,j);
      }
   }
   mean = mean / (float)(N*N);
   output3 << t << "\t" << mean << endl;

   float rough = rms(mean);
   output2 << t << "\t" << rough << endl;

   if(t%100 == 0) //Output surface profile
   {
      ofstream output;
```

```
            char fname[10];
            itoa(t,fname,10);
            output.open(string(filenm+string(" ")+
            string(fname)+string(".txt")).c_str());

            for(int i = 0; i < N; i++)
            {
               for(int j = 0; j < N; j++)
               {
                  output << h(i,j) << " ";
               }
               output << endl;
            }
            output.close();
         }
      }

      return 0;
   }
```

D

Solid-on-Solid Model Implementation

The following is a C++ implementation of the solid-on-solid model discussed in Sect. 8.1. This code models a deposition with normally incident particle flux and surface diffusion.

```
#include <iostream>
#include <fstream>
#include <string>
#include <cmath>

using namespace std;

const int N = 32768;
//Lattice size is N, must be less than or = (RAND_MAX - 1)
int h[N]; //Surface height array
double T = 1e8; // Number of particles to deposit

int jumps = 10; //Number of diffusion jumps

float mean = 0; //Mean surface height
float rough = 0; //Surface roughness

double Ea = 0.08; //Activation energy in eV
double kb = 8.6174e-5; //Boltzmann constant in eV/K
double Ts = 350; //Temperature of substrate in K
double En = 0.05; //Nearest neighbor activation energy

int height(int x)
//Returns surface height
{
    while(x < 0){ x += N;}
    while(x>(N-1)){ x -= N;}
```

```
    return h[x];
}

void deposit(int x)
//Increases surface height
{
    while(x < 0){ x += N;}
    while(x>(N-1)){ x -= N;}

    h[x]++;
}

void diffuse(int x, int xf)
//Diffuse particle from x to xf
{
    while(x < 0){ x += N;}
    while(x>(N-1)){ x -= N;}

    while(xf < 0){ xf += N;}
    while(xf>(N-1)){ xf -= N;}

    h[x]--;
    h[xf]++;
}

void evolve() //Add a new particle
{
     //Randomly choose initial deposition point
     //uniformly in [0,N-1]
     int x = -1;
     while (x<0 || x >(N-1)){ x = (int)( rand() );}

     deposit(x);

     //Particle diffusion
     for(int k = 0; k < jumps; k++)
     {
         //Choose particle to diffuse
         int l = -1;

          while (l<0 || l >(N-1)){ l = (int)( rand() );}
          int xd = l;
          if(height(xd)==0) continue;
          //Cannot diffuse if no particles
```

```
        int nn = 0; //Nearest neighbors
        //Count nearest neighbors
        for(int a = -1; a <=1; a++)
        {
             if(height(xd) <= height(xd+a)) nn++;
        }

       //Attempt to diffuse if particle has enough energy
       while((float)rand()/(RAND_MAX)
              < exp(-(Ea+nn*En)/(kb*Ts) ))
       {
           //Choose diffusion site
          int i = 0;
          while(i==0 || i < -1 || i > 1)
          {i = (int)( 2*((double)rand()/(RAND_MAX)
                      -0.5) * (2) );}

          //If diffusion site is no more
          //than 1 taller, diffuse
          int delta_h = (int) ( height(xd+i)-height(xd) );
          float delta_r = (float)
                           sqrt( delta_h*delta_h+ i*i );
          if (delta_h<=0 || delta_r<2 )
          {
               diffuse(xd,xd+i);
               xd = xd + i;
          }

          //Continue diffusing until Boltzmann factor
          //stops diffusion
          nn = 0;
          for(int a = -1; a <=1; a++)
          {
               if(height(xd) <= height(xd+a)) nn++;
          }
       }
   }
}

int main()
{
    //Initialize random numbers
    srand( (unsigned)time( NULL ) );
```

```
ofstream output; //Output file streams
ofstream surface;
ofstream acf;

string filenm = "jump10";
//Filename for output files

output.open(string(filenm
                   +string("_stats.txt")).c_str());
surface.open(string(filenm
                    +string("_surface.txt")).c_str());

for(int i = 0; i < N; i++)
{
       h[i] = 0; //Flat initial surface
}

for(int t = 1; t <= T; t++)
{
      evolve(); //Add a new particle

      if(t%100000 == 0)
      {
         //Find mean height
         mean = 0;
         for(int i = 0; i < N; i++)
         {
            mean += h[i];
         }

         mean = (float) mean / (N);

         //Find roughness
         rough = 0;
         for(int i = 0; i < N; i++)
         {
            rough += (h[i] - mean)*(h[i] - mean);
         }

         rough = sqrt(rough / (N) );

         //Output mean height and roughness
         output << t << " " << mean << " "
                  << rough << endl;
```

```
            if(t%10000000 == 0)
            //Output autocorrelation function
            {
               char fname[15];
               itoa(t,fname,10);

               acf.open(string(filenm+string("_acf_")+
                string(fname)+string(".txt")).c_str());
               for(int dx = 0; dx < N/2; dx++)
               {
                  float sum = 0;
                  for(int i = 0; i < N; i++)
                  {
                     sum+= (height(i+dx)- mean)
                            *(height(i)-mean);
                  }

                    acf << dx<< " " << sum/(N) << endl;
               }
               acf.close();
            }

          }
     }

     for(int i = 0; i < N; i++) //Output surface profile
     {
            surface << i << " " <<(h[i] - mean)<<endl;
     }

   return 0;
}
```

References

1. J.G. Amar, F. Family, P.-M. Lam: Dynamic scaling of the island-size distribution and percolation in a model of submonolayer molecular-beam epitaxy. Phys. Rev. B **50**, 8781 (1994)
2. J.G. Amar, P.-M. Lam, F. Family: Groove instabilities in surface growth with diffusion. Phys. Rev. E **47**, 3242 (1993)
3. J.G. Amar, P.-M. Lam, F. Family: Surface growth with long-range correlated noise. Phys. Rev. A **43**, 4548 (1991)
4. G. Arfken, H. Weber: *Mathematical Methods for Physicists,* 5th edn (Harcourt Academic Press, New York 2001) p 711
5. P.B. Balbuena, J. M. Seminario: *Molecular Dynamics: From Classical to Quantum Methods* (Elsevier, New York 1999)
6. G.S. Bales, A.C. Redfield, A. Zangwill: Growth dynamics of chemical vapor deposition. Phys. Rev. Lett. **62**, 776 (1989)
7. G.S. Bales, A. Zangwill: Growth dynamics of sputter deposition. Phys. Rev. Lett. **63**, 692 (1989)
8. A.-L. Barabási, H.E. Stanley: *Fractal Concepts in Surface Growth* (Cambridge University Press, New York 1995)
9. A.-L. Barabási, H.E. Stanley: *Fractal Concepts in Surface Growth* (Cambridge University Press, New York 1995) p 19
10. A.-L. Barabási, H.E. Stanley: *Fractal Concepts in Surface Growth* (Cambridge University Press, New York 1995) p 27
11. A.-L. Barabási, H.E. Stanley: *Fractal Concepts in Surface Growth* (Cambridge University Press, New York 1995) p 213
12. A.-L. Barabási, H.E. Stanley: *Fractal Concepts in Surface Growth* (Cambridge University Press, New York 1995) p 289
13. D. Bensimon, B. Shraiman, S. Liang: On the ballistic model of aggregation. Physics Lett. A **102**, 238 (1984)
14. G.E.P. Box, M.E. Muller: A note on the generation of random normal deviates. Ann. Math. Stat. **29**, 610 (1958)
15. M. Boas: *Mathematical Methods in the Physical Sciences,* 2nd edn (John Wiley & Sons, New Jersey 1983) p 519
16. R. Bracewell: *The Fourier Transform and Its Applications,* 3rd edn (McGraw-Hill, New York 1999) pp 69–97

17. P. Brault, P. Dumas, F. Salvan: Roughness scaling of plasma-etched silicon surfaces. J. Phys.: Condens. Matter **10**, L27 (1998)
18. C. Buzea, G. Beydaghyan, C. Elliott, K. Robbie: Control of power law scaling in the growth of silicon nanocolumn pseudo-regular arrays deposited by glancing angle deposition. Nanotechnology **16**, 1986 (2005)
19. A.-C.-K. Chan, H. Wang, M.J. Chan: High quality thermal oxide on LPSOI formed by high temperature enhanced MILC. IEEE Electron Device Letters **22**, 384 (2001)
20. R. Chiarello, V. Panella, J. Krim, C. Thompson: X-ray reflectivity and adsorption isotherm study of fractal scaling in vapor-deposited films. Phys. Rev. Lett. **67**, 3408 (1991)
21. M. Constantin, C. Dasgupta, P.P. Chatraphorn, S.N. Majumdar, S. Das Sarma: Persistence in nonequilibrium surface growth. Phys. Rev. E **69** 061608 (2004)
22. Image courtesy of George T. Dalakos. G.T. Dalakos, J.P. Plawsky, P.D. Persans: Topographic evolution during deposition of plasma-deposited hydrogenated silicon on glass. Phys. Rev. B **72**, 205305 (2005)
23. S. Das Sarma, S.V. Ghaisas, J.M. Kim: Kinetic super-roughening and anomalous dynamic scaling in nonequilibrium growth models. Phys. Rev. E **49**, 122 (1994)
24. S. Das Sarma, C.J. Lanczycki, R. Kotlyar, S.V. Ghaisas: Scale invariance and dynamical correlations in growth models of molecular beam epitaxy. Phys. Rev. E **53**, 359 (1996)
25. S. Das Sarma, P. Tamborenea: A new universality class for kinetic growth: One-dimensional molecular-beam epitaxy. Phys. Rev. Lett. **66**, 325 (1991)
26. P. Dennery, A. Krzywicki: *Mathematics for Physicists* (Dover, New York 1995)
27. L. Devroye: *Non-Uniform Random Variate Generation* (Springer-Verlag, New York 1986)
28. C.V. Dharmadhikari, A.O. Ali, N. Suresh, D.M. Phase, S.M. Chaudhari, V. Ganesan, A. Gupta, B.A. Dasannacharya: Dynamic scaling in growth of platinum films on Si(100). Mater. Sci. Eng. B **75**, 29 (2000)
29. B. Dick, M.J. Brett, T. Smy, M. Belov, M.R. Freeman: Periodic submicrometer structures by sputtering. J. Vac. Sci. Technol. B **19**, 1813 (2001)
30. B. Dick, M.J. Brett, T.J. Smy, M.R. Freeman, M. Malac, R.F. Egerton: Periodic magnetic microstructures by glancing angle deposition. J. Vac. Sci. Technol. A **18**, 1838 (2000)
31. D. M. Dobkin, M. K. Zuraw, eds: *Principles of Chemical Vapor Deposition* (Springer, New York 2006)
32. Reprinted with permission from: J.T. Drotar, Y.-P. Zhao, T.-M. Lu, G.-C. Wang: Mechanisms for plasma and reactive ion etch-front roughening. Phys. Rev. B **61**, 3012 (2000); Copyright 2000 by the American Physical Society
33. J.T. Drotar, Y.-P. Zhao, T.-M. Lu, G.-C. Wang: Numerical analysis of the noisy Kuramoto-Sivashinsky equation in 2+1 dimensions. Phys. Rev. E **59**, 177 (1999)
34. J.T. Drotar, Y.-P. Zhao, T.-M. Lu, G.-C. Wang: Surface roughening in shadowing growth and etching in 2+1 dimensions. Phys. Rev. B **62**, 2118 (2000)
35. J.T. Drotar, Y.-P. Zhao, T.-M. Lu, G.-C. Wang: Surface roughening in low-pressure chemical vapor deposition. Phys. Rev. B **64**, 125411 (2001)
36. J.T. Drotar, Y.-P. Zhao, T.-M. Lu, G.-C. Wang: Why is KPZ type surface roughening so hard to observe?. Mat. Res. Soc. Proc. **648**, 7.9.1 (2001)

37. L. Dubost, A. Rhallabi, J. Perrin, J. Schmitt: Growth of nodular defects during film deposition. J. Appl. Phys. **78**, 3784 (1995)
38. S.F. Edwards, D.R. Wilkinson: The surface statistics of a granular aggregate. Proc. R. Soc. London Ser. A **381**, 17 (1982)
39. F. Family: Scaling of rough surfaces: Effects of surface diffusion. J. Phys. A: Math. Gen. **19**, L441 (1986)
40. F. Family, T. Vicsek: *Dynamics of Fractal Surfaces* (World Scientific Publishing Co., New Jersey 1991)
41. F. Family, T. Vicsek: Scaling of the active zone in the Eden process on percolation networks and the ballistic deposition model. J. Phys. A: Math. Gen. **18**, L75 (1985)
42. G. Fisherman: *Monte Carlo: Concepts, Algorithms, and Applications* (Springer-Verlag, New York 1995)
43. H. Guclu, G. Korniss: Extreme fluctuations in small-world networks with relaxational dynamics. Phys. Rev. E **69**, 065104 (2004)
44. K.H. Guenther: Microstructure of vapor-deposited optical coatings. Appl. Opt. **23**, 3806 (1984)
45. S. Hassani: *Mathematical Physics* (Springer-Verlag, New York 1999)
46. M.M. Hawkeye, M.J. Brett: Narrow bandpass optical filters fabricated with one-dimensionally periodic inhomogeneous thin films. J. Appl. Phys. **100**, 044322 (2006)
47. P.E. Hegeman, H.J.W. Zandvliet, G.A.M. Kip, A. van Silfhout: Kinetic roughening of vicinal Si(001). Surf. Sci. **311**, L655 (1994)
48. B M. Henry, Z. Xie, H. Smith, K.R. Kirov, V.M. Burlakov, C.R.M. Grovenor, H.E. Assender, G.A.D. Briggs, P.L. Burn, M. Kano, Y. Tsukahara: Characterization and performance of TiO-based solar cells. Proc. SPIE **5938**, 593817 (2005)
49. I. Hodgkinson, Q.H. Wu, S. Collett: Dispersion equations for vacuum-deposited tilted-columnar biaxial media. Appl. Opt. **40**, 452 (2001)
50. I. Hodgkinson, Q.H. Wu, J. Hazel: Empirical equations for the principal refractive indices and column angle of obliquely deposited films of tantalum oxide, titanium oxide, and zirconium oxide. Appl. Opt. **37**, 2653 (1998)
51. B. Holley, A. Faghri: Analysis of pulsating heat pipe with capillary wick and varying channel diameter. Int. J. of Heat and Mass Trans. **48**, 2635 (2005)
52. P. Hrudey, A. Popta, J. Sit, M. Brett: Photonic device applications of nano-engineered thin film materials. Proc. SPIE **5931**, 593113 (2005)
53. P. Jain, J.S. Juneja, T. Karabacak, E.J. Rymaszewski, T.-M. Lu: Surface roughness evolution in amorphous tantalum oxide films deposited by pulsed reactive sputtering. In: *Morphological and Compositional Evolution of Thin Films*, ed by M.J. Aziz, N.C. Bartelt, I. Berbezier, J.B. Hannon, S.J. Hearne. Mat. Res. Soc. Proc. **749**, W5.11 (2003)
54. J.H. Jeffries, J.-K. Zuo, M.M. Craig: Instability of kinetic roughening in sputter-deposition growth of Pt on glass. Phys. Rev. Lett. **76**, 4931 (1996)
55. M.D. Johnson, C. Orme, A.W. Hunt, D. Graff, J. Sudijono, L.M. Sander, B.G. Orr: Stable and unstable growth in molecular beam epitaxy. Phys. Rev. Lett. **72**, 116 (1994)
56. K. Kaminska, A. Amassian, L. Martinu, K. Robbie: Growth of vacuum evaporated ultraporous silicon studied with spectroscopic ellipsometry and scanning electron microscopy. J. Appl. Phys. **97**, 013511 (2005)

57. T. Karabacak, T.-M. Lu: Shadowing growth and physical self-assembly. In: *Handbook of Theoretical and Computational Nanotechnology,* ed by M. Rieth, W. Schommers (American Scientific Publishers, California 2005) ch 69
58. Reprinted with permission from: T. Karabacak, J.P. Singh, Y.-P. Zhao, G.-C. Wang, T.-M. Lu: Scaling during shadowing growth of isolated nanocolumns. Phys. Rev. B **68**, 125408 (2003); Copyright 2003 by the American Physical Society
59. Reprinted with permission from: T. Karabacak, Y.-P. Zhao, G.-C. Wang, T.-M. Lu: Growth-front roughening in amorphous silicon films by sputtering. Phys. Rev. B **64**, 085323 (2001); Copyright 2001 by the American Physical Society
60. T. Karabacak, Y.-P. Zhao, G.-C. Wang, T.-M. Lu: Growth front roughening in silicon nitride films by plasma-enhanced chemical vapor deposition. Phys. Rev. B **66**, 075329 (2002)
61. M. Kardar, G. Parisi, Y.C. Zhang: Dynamic scaling of growing interfaces. Phys. Rev. Lett. **56**, 889 (1986)
62. R.-P.-U. Karunasiri, R. Bruinsma, J. Rudnick: Thin-film growth and the shadow instability. Phys. Rev. Lett. **62**, 788 (1989)
63. E. Katzav: Self-consistent expansion for the molecular beam epitaxy equation. Phys. Rev. E **65**, 032103 (2002)
64. P. Keblinski, A. Maritan, F. Toigo, R. Messier, J.R. Banavar, Continuum model for the growth of interfaces. Phys. Rev. E **53**, 759 (1996)
65. S.R. Kennedy, M.J. Brett, H. Miguez, O. Toader, S. John: Optical properties of a three-dimensional silicon square spiral photonic crystal. Photonics and Nanostruct. Fundam. and Appl. **1**, 37 (2003)
66. S. Kennedy, M.J. Brett, O. Toader, S. John: Fabrication of tetragonal square spiral photonic crystals. Nano Lett. **2**, 59 (2001)
67. S.V. Kesapragada, D. Gall: Anisotropic broadening of Cu nanorods during glancing angle deposition. Appl. Phys. Lett. **89**, 203121 (2006)
68. J.M. Kim, S. Das Sarma: Discrete models for conserved growth equations. Phys. Rev. Lett. **72**, 2903 (1994)
69. S. Kim, N. Koratkar, T. Karabacak, T.-M. Lu: Water electrolysis activated by Ru nanorod array electrodes. Appl. Phys. Lett. **88**, 263106 (2006)
70. B. Kozma, M.B. Hastings, G. Korniss: Diffusion processes on power-law small-world networks. Phys. Rev. Lett. **95**, 018701 (2005)
71. B. Kozma, M.B. Hastings, G. Korniss: Roughness scaling for Edwards-Wilkinson relaxation in small-world networks. Phys. Rev. Lett. **92**, 108701 (2004)
72. J. Krug, P. Meakin: Columnar growth in oblique incidence ballistic deposition: Faceting, noise reduction, and mean-field theory. Phys. Rev. A **43**, 900 (1991)
73. Y. Kuo: *Thin Film Transistors: Materials and Processes* (Springer-Verlag, New York 2003).
74. Z.W. Lai, S. Das Sarma: Kinetic growth with surface relaxation: Continuum versus atomistic models. Phys. Rev. Lett. **66**, 2348 (1991)
75. A. Lakhtakia: Sculptured thin films: Accomplishments and emerging uses. Mater. Sci. Eng. C **19**, 427 (2002)
76. A. Lakhtakia, R. Messier: *Sculptured Thin Films: Nanoengineered Morphology and Optics,* (SPIE Press Monograph vol PM143, 2005).
77. C.-H. Lam, L.M. Sander: Surface growth with power-law noise. Phys. Rev. Lett. **69**, 3338 (1992)

78. P.-M. Lam, F. Family: Dynamics of a height-conserved surface-growth model with spatially correlated noise. Phys. Rev. A **44**, 7939 (1991)
79. Z. Lang, W. Xiuqin: Formation of nodular defects as revealed by simulation of a modified ballistic model of depositional growth. J. Mat. Sci. **33**, 1487 (1998)
80. H. Leschhorn, L.-H. Tang: Comment on 'Elastic string in a random potential'. Phys. Rev. Lett. **70**, 2973 (1993)
81. S. Liang, L.P. Kadanoff: Scaling in a ballistic aggregation model. Phys. Rev. A **31**, 2628 (1985)
82. C.-H. Lin, F. Yuan, C.-R. Shie, K.-F. Chen, B.-C. Hsu, M.H. Lee, W.W. Pai, C.W. Liu: Roughness-enhanced reliability of MOS tunneling diodes. IEEE Electron Device Letters **23**, 431 (2002)
83. A.V. Limaye, R.E. Amritkar: Theory of growth of ballistic aggregates. Phys. Rev. A **34**, 5085 (1986)
84. A. Lisfi, J.C. Lodder: Magnetic domains in Co thin films obliquely sputtered on a polymer substrate. Phys. Rev. B **63**, 174441 (2001)
85. F. Liu, M.T. Umlor, L. Shen, J. Weston, W. Eads, J.A. Barnard, G.J. Mankey: The growth of nanoscale structured iron films by glancing angle deposition. J. Appl. Phys. **85**, 5486 (1999)
86. G.P. Lepage: A new algorithm for adaptive multidimensional integration. J. Comp. Phys. **27**, 192 (1978)
87. J.M. López: Scaling approach to calculate critical exponents in anomalous surface roughening. Phys. Rev. Lett. **83**, 4594 (1999)
88. J.M. López, M.A. Rodríguez: Lack of self-affinity and anomalous roughening in growth processes. Phys. Rev. E **54**, R2189 (1996)
89. J.M. López, M.A. Rodríguez: Growth of interfaces with strong quenched disorder: Columnar media. Phys. Rev. E **52**, 6442 (1995)
90. J.M. López, M.A. Rodríguez, R. Cuerno: Superroughening versus intrinsic anomalous scaling of surfaces. Phys. Rev. E **56**, 3993 (1997)
91. T.-M. Lu, H.-N. Yang, G.-C. Wang: When interface gets rough.... Mat. Res. Soc. Symp. Proc. **367**, 283 (1995)
92. For review, see T.-M. Lu, Y.-P. Zhao, J.T. Drotar, T. Karabacak, G.-C. Wang: Novel mechanisms on the growth morphology of films. In: *Morphological and Compositional Evolution of Thin Films*, ed by M.J. Aziz, N.C. Bartelt, I. Berbezier, J.B. Hannon, S.J. Hearne. Mat. Res. Soc. Proc. **749**, W1.2 (2003)
93. D.R. Luhman, R.B. Hallock: Evidence for power-law dominated noise in vacuum deposited CaF_2. Phys. Rev. Lett. **92**, 256102 (2004)
94. M. Lutt, J.P. Schlomka, M. Tolan, J. Stettner, O.H. Seeck, W. Press: Kardar-Parisi-Zhang growth of amorphous silicon on Si/SiO_2. Phys. Rev. B **56**, 4085 (1997)
95. J. Mahan: *Physical Vapor Deposition of Thin Films* (John Wiley & Sons, Inc., New York 2000)
96. J. Mahan: *Physical Vapor Deposition of Thin Films* (John Wiley & Sons, Inc., New York 2000) p 29
97. J. Mahan: *Physical Vapor Deposition of Thin Films* (John Wiley & Sons, Inc., New York 2000) p 153
98. E. Main, T. Karabacak, T.-M. Lu: Continuum model for nanocolumn growth during oblique angle deposition. J. Appl. Phys. **95**, 4346 (2004)
99. S. Majaniemi, T. Ala-Nissila, J. Krug: Kinetic roughening of surfaces: Derivation, solution, and application of linear growth equations. Phys. Rev. B **53**, 8071 (1996)

100. B.B. Mandelbrot: Self-affine fractals and fractal dimension. Physica Scripta **32**, 257 (1985)
101. A. Margolina, H.E. Warriner: Growth in a restricted solid on solid model with correlated noise. J. Stat. Phys. **60**, 809 (1990)
102. J. Mathews, R. L. Walker: *Mathematical Methods of Physics,* 2nd edn (Addison-Wesley, New York 1970)
103. S.G. Mayr, K. Samwer: Tailoring the surface morphology of amorphous thin films by appropriately chosen deposition conditions. J. Appl. Phys. **91**, 2779 (2002)
104. P. Meakin: *Fractals, Scaling, and Growth Far from Equilibrium* (Cambridge University Press, New York 1998)
105. P. Meakin: *Fractals, Scaling, and Growth Far from Equilibrium* (Cambridge University Press, New York 1998) p 420
106. P. Meakin: *Fractals, Scaling, and Growth Far from Equilibrium* (Cambridge University Press, New York 1998) p 553
107. P. Meakin, J. Krug: Scaling structure in simple screening models for columnar growth. Phys. Rev. A **46**, 4654 (1992)
108. P. Meakin, R. Jullien: Restructuring effect in the rain model for random deposition. J. Physique **48**, 1651 (1987)
109. P. Meakin, R. Jullien: Spatially correlated ballistic deposition on one- and two-dimensional surfaces. Phys. Rev. A **41**, 983 (1990)
110. P. Meakin, R. Jullien: Spatially correlated ballistic deposition on one- and two-dimensional surfaces. Europhys. Lett. **9**, 71 (1989)
111. P. Meakin, P. Ramanlal, L.M. Sander, R.C. Ball: Ballistic deposition on surfaces. Phys. Rev. A **34**, 5091 (1986)
112. T. Michely, J. Krug: *Islands, Mounds, and Atoms: Patterns and Processes in Crystal Growth Far from Equilibrium* (Springer-Verlag, New York 2003)
113. D. Moldovan, L. Golubovic: Interfacial coarsening dynamics in epitaxial growth with slope selection. Phys. Rev. E **61**, 6190 (2000)
114. W.W. Mullins: Theory of thermal grooving. J. Appl. Phys. **28**, 333 (1957)
115. V. Ng, J.F. Hu, A.O. Adeyeye, J.P. Wang, T.C. Chong: Factors affecting surface roughness and coercivity of $Ni_{80}Fe_{20}$ thin films. J. Appl. Phys. **91**, 7206 (2002)
116. F. Ojeda, R. Cuerno, R. Salvarezza, L. Vázquez: Dynamics of rough interfaces in chemical vapor deposition: Experiments and a model for silica films. Phys. Rev. Lett. **84**, 3125 (2000)
117. S. Pal, D.P. Landau: Monte Carlo simulation and dynamic scaling of surfaces in MBE growth. Phys. Rev. B **49**, 10597 (1994)
118. G. Palasantzas: Roughness spectrum and surface width of self-affine fractal surfaces via the K-correlation model. Phys. Rev. B **48**, 14472 (1993)
119. G. Palasantzas: Self-affine fractals and the limit $H \to 0$. Phys. Rev. E **49**, 1740 (1994)
120. G. Palasantzas, J. Krim: Effect of the form of the height-height correlation function on diffuse x-ray scattering from a self-affine surface. Phys. Rev. B **48**, 2873 (1993)
121. A. Papoulis, *Probability, Random Variables, and Stochastic Processes,* 2nd edn (McGraw-Hill, New York 1984) pp 102–103
122. Reprinted with permission from: M. Pelliccione, T. Karabacak, C. Gaire, G.-C. Wang, T.-M. Lu: Mound formation in surface growth under shadowing. Phys. Rev. B **74**, 125420 (2006); Copyright 2006 by the American Physical Society

123. M. Pelliccione, T. Karabacak, T.-M. Lu: Breakdown of dynamic scaling in surface growth under shadowing. Phys. Rev. Lett. **96**, 146105 (2006)
124. Reprinted with permission from: M. Pelliccione, T.-M. Lu: Self-shadowing in ballistic fan formation from point seeds. Phys. Rev. B **75**, 245431 (2007); Copyright 2007 by the American Physical Society
125. C.-K. Peng, S. Havlin, M. Schwartz, H.E. Stanley: Directed polymer and ballistic-deposition growth with correlated noise. Phys. Rev. A **44**, R2239 (1991)
126. R. Pétri, P. Brault, O. Vatel, D. Henry, E. André, P. Dumas, F. Salvan: Silicon roughness induced by plasma etching, J. Appl. Phys. **75**, 7498 (1994)
127. J. Poortmans, V. Arkhipov, eds: *Thin Film Solar Cells: Fabrication, Characterization and Applications* (John Wiley & Sons, New York 2006)
128. F. Porcú, F. Prodi: Ballistic accretion on seeds of different sizes. Phys. Rev. A **44**, 8313 (1991)
129. W.H. Press, S.A. Teukolsky, W.T. Vetterling, B.P. Flannery, *Numerical Recipes: The Art of Scientific Computing,* 3rd edn (Cambridge University Press, New York 2007)
130. R. Rabady, I. Avrutsky: Reduced surface roughness of solid thin films prepared by alternating-bias, radio-frequency magnetron sputtering. J. Opt. Soc. Am. B **20**, 2174 (2003)
131. M. Raible, S.J. Linz, P. Hänggi: Amorphous thin film growth: Minimal deposition equation. Phys. Rev. E **62**, 1691 (2000)
132. P. Ramanlal, L.M. Sander: Theory of ballistic aggregation. Phys. Rev. Lett. **54**, 1828 (1985)
133. D.C. Rapaport: *The Art of Molecular Dynamics Simulation* (Cambridge University Press, New York 2004)
134. D. Reicher, P. Black, K. Jungling: Defect formation in hafnium dioxide thin films. Appl. Opt. **39**, 1589 (2000)
135. S.M. Rossnagel, T.S. Kuan: Alteration of Cu conductivity in the size effect regime. J. Vac. Sci. Technol. B **22**, 240 (2004)
136. P.I. Rovira, R.A. Yarussi, R.W. Collins, V.C. Venugopal, A. Lakhtakia, R. Messier, K. Robbie, M.J. Brett: Rotating-compensator multichannel transmission ellipsometry of a thin-film helicoidal bianisotropic medium. Thin Solid Films **313**, 373 (1998)
137. R.A. Roy, R.Messier: Preparation-physical structure relations in SiC sputtered films. J. Vac. Sci. Tech. A **2**, 312 (1984)
138. M. Saitou: Anomalous scaling of nickel surfaces in pulse-current electrodeposition growth. Phys. Rev. B **66**, 073416 (2002)
139. J. Shao, S. Wang, Z. Shen, X. Fu, H. He, Z. Fan: Glancing angle deposited thin films and their applications in laser systems. Proc. SPIE **6403**, 640312 (2006)
140. M. Siegert: Coarsening dynamics of crystalline thin films. Phys. Rev. Lett. **81**, 5481 (1998)
141. M. Siegert, M. Plischke: Slope selection and coarsening in molecular beam epitaxy. Phys. Rev. Lett. **73**, 1517 (1994)
142. S.K. Sinha, E.B. Sirota, S. Garoff, H.B. Stanley: X-ray and neutron scattering from rough surfaces. Phys. Rev. B **38**, 2297 (1988)
143. A.H.M. Smets, W.M.M. Kessels, M.C.M. van de Sanden: Temperature dependence of the surface roughness evolution during hydrogenated amorphous silicon film growth. Appl. Phys. Lett. **82**, 865 (2003)

144. D.L. Smith: *Thin Film Deposition: Principles and Practice* (McGraw-Hill, New York 1995) p 307
145. W.A. Smith: *Elementary Numerical Analysis* (Prentice Hall, New Jersey 1986) p 21
146. I.M. Sokolov, J. Mai, A. Blumen: Paradoxal diffusion in chemical space for nearest-neighbor walks over polymer chains. Phys. Rev. Lett. **79**, 857 (1997)
147. D. Sornette: *Critical Phenomena in Natural Sciences: Chaos, Fractals, Selforganization and Disorder: Concepts and Tools* (Springer, New York 2004) p 254
148. B.A. Sperling, J.R. Abelson: Kinetic roughening of amorphous silicon during hot-wire chemical vapor deposition at low temperature. J. Appl. Phys. **101**, 024915 (2007) and references therein
149. M.R. Spiegel, J. Liu: *Schaum's Outlines: Mathematical Handbook of Formulas and Tables,* 2nd edn (McGraw-Hill, New York 1999) p 109 eqn 18.71
150. H.E. Stanley: *Introduction to Phase Transition and Critical Phenomena* (Oxford University Press, New York 1971) p 108
151. C. Stolz, R. Tench, M. Kozlowski, A. Fornier: A comparison of nodular defect seed geometries from different deposition techniques. Proc. SPIE **2714**, 374 (1996)
152. J.A. Stroscio, D.T. Pierce, M.D. Stiles, A. Zangwill, L.M. Sander: Coarsening of unstable surface features during Fe(001) homoepitaxy. Phys. Rev. Lett. **75**, 4246 (1995)
153. S.M. Sze, K.K. Ng: *Physics of Semiconductor Devices,* 3rd edn (John Wiley & Sons, New York 2006)
154. D.M. Tanenbaum, A.L. Laracuente, A. Gallagher: Surface roughening during plasma-enhanced chemical-vapor deposition of hydrogenated amorphous silicon on crystal silicon substrates. Phys. Rev. B **56**, 4243 (1997)
155. C. Tang, S. Alexander, R. Bruinsma, B.E. Shaw: Scaling theory for the growth of amorphous films. Phys. Rev. Lett. **64**, 772 (1990)
156. F. Tang, T. Karabacak, L. Li, M. Pelliccione, G.-C. Wang, T.-M. Lu: Power-law scaling during shadowing growth of nanocolumns by oblique angle deposition. J. Vac. Sci. Tech. A **25**, 160 (2007)
157. F. Tang, D.-L. Liu, D.-X. Ye, Y.-P. Zhao, T.-M. Lu, A. Vijayaraghavan: Magnetic properties of Co nanocolumns fabricated by oblique-angle deposition. J. Appl. Phys. **93**, 4194 (2003)
158. R. Teki, N. Koratkar, T. Karabacak, T.-M. Lu: Enhanced photoemission from nanostructured surface topologies. Appl. Phys. Lett. **89**, 193116 (2006)
159. J.A. Thornton: High rate thick film growth. Annu. Rev. Mater. Sci. **7**, 239 (1977)
160. E.S. Tok, S.W. Ong, H.C. Kang: Dynamical scaling of sputter-roughened surfaces in 2+1 dimensions. Phys. Rev. E **70**, 011604 (2004)
161. F. Tsui, J. Wellman, C. Uher, R. Clarke: Morphology transition and layer-by-layer growth of Rh(111). Phys. Rev. Lett. **76**, 3164 (1996)
162. J.E. Van Nostrand, S.J. Chey, M.-A. Hasan, D.G. Cahill, J.E. Greene: Surface morphology during multilayer epitaxial growth of Ge(001). Phys. Rev. Lett. **74**, 1127 (1995)
163. J.A. Venables: *Introduction to Surface and Thin Film Processes* (Cambridge University Press, New York 2000) p 144
164. D. Vick, T. Smy, B. Dick, S. Kennedy, M.J. Brett: Growth behavior of engineered porous thin films - Measurement and modeling. Mat. Res. Soc. Symp. Proc. **648**, 3.43.1 (2001)

165. J. Villain: Continuum models of crystal growth from atomic beams with and without deposition. J. Phys. I (France) **1**, 19 (1991)
166. For review, see E.G. Wang: *Handbook of Theoretical and Computational Nanotechnology,* ed by M. Rieth, W. Schommers (American Scientific, California 2006) p 256
167. G.-C. Wang, T.-M. Lu: Physical realization of two-dimensional Ising critical phenomenon: Oxygen chemisorbed on a W(112) surface. Phys. Rev. **B** 31, 5918 (1985)
168. G.N. Watson: *A Treatise on the Theory of Bessel Functions* (Cambridge University Press, New York 1944) p 410
169. D.J. Watts: *Small Worlds: The Dynamics of Networks Between Order and Randomness* (Princeton University Press, New Jersey 1999)
170. D.J. Watts, S.H. Strogatz: Collective dynamics of 'small-world' networks. Nature (London) **393**, 440 (1998)
171. T.A. Witten Jr, L.M. Sander: Diffusion-limited aggregation, a kinetic critical phenomenon. Phys. Rev. Lett. **47**, 1400 (1981)
172. D.E. Wolf, J. Villain: Growth with surface diffusion. Europhys. Lett. **13**, 389 (1990)
173. J.-Q. Xi, J.K. Kim, E.F. Schubert, D. Ye, T.-M. Lu, S.-Y. Lin, J.S. Juneja: Very low-refractive-index optical thin films consisting of an array of SiO_2 nanorods. Opt. Lett. **31**, 601 (2006)
174. Reprinted with permission from: H.-N. Yang, Y.-P. Zhao, A. Chan, T.-M. Lu, G.-C. Wang: Sampling-induced hidden cycles in correlated random rough surfaces. Phys. Rev. B **56**, 4224 (1997); Copyright 1997 by the American Physical Society
175. H.-N. Yang, Y.-P. Zhao, G.-C. Wang, T.-M. Lu: Noise-induced roughening evolution of amorphous Si films grown by thermal evaporation. Phys. Rev. Lett. **76**, 3774 (1996)
176. A. Yanguas-Gil, J. Cotrino, A. Barranco, A.R. González-Elipe Influence of the angular distribution function of incident particles on the microstructure and anomalous scaling behavior of thin films. Phys. Rev. Lett. **96**, 236101 (2006)
177. J.H. Yao, H. Guo: Shadowing instability in three dimensions. Phys. Rev. E **47**, 1007 (1993)
178. J. Yao, C. Roland, H. Guo: Interfacial dynamics with long-range screening. Phys. Rev. A **45**, 3903 (1992)
179. D.-X. Ye, T.-M. Lu: Fanlike aggregations on seeds by parallel ballistic flux: Experimental results and Monte Carlo simulations of the growth of three-dimensional Si structures. Phys. Rev. B **75**, 115420 (2007)
180. D.-X. Ye, Z.-P. Yang, A.S.P. Chang, J. Bur, S.-Y. Lin, T.-M. Lu, R.Z. Wang, S. John: Experimental realization of a well-controlled 3D silicon spiral photonic crystal. J. Phys. D: Appl. Phys. **40**, 2624 (2007)
181. J. Yu, J.G. Amar: Dynamical scaling behavior in two-dimensional ballistic deposition with shadowing. Phys. Rev. E **66**, 021603 (2002)
182. Y.-C. Zhang: Growth anomaly and its implications. Physica A **170**, 1 (1990)
183. Y.-C. Zhang: Non-universal roughening of kinetic self-affine surfaces. J. Physique **51**, 2129 (1990)
184. Y.-P. Zhao, J.T. Drotar, G.-C. Wang, T.-M. Lu: Morphology transition during low-pressure chemical vapor deposition. Phys. Rev. Lett. **87**, 136102 (2001)
185. Y.-P. Zhao, J.T. Drotar, G.-C. Wang, T.-M. Lu: Roughening in plasma etch fronts of Si(100). Phys. Rev. Lett. **82**, 4882 (1999)

186. Y.-P. Zhao, J.B. Fortin, G. Bonvallet, G.-C. Wang, T.-M. Lu: Kinetic roughening in polymer film growth by vapor deposition. Phys. Rev. Lett. **85**, 3229 (2000)
187. Y.-P. Zhao, G.-C. Wang, T.-M. Lu: *Characterization of Amorphous and Crystalline Rough Surface: Principles and Applications* (Academic Press, New York 2001)
188. Y.-P. Zhao, D.-X. Ye, P.-I. Wang, G.-C. Wang, T.-M. Lu: Fabrication of Si nanocolumns and Si square spirals of self-assembled monolayer colloid substrates. Int. J. Nanosci. **1**, 87 (2002)
189. C.M. Zhou, D. Gall: Competitive growth of Ta nanopillars during glancing angle deposition: Effect of surface diffusion. J. Vac. Sci. Technol. A **25**, 312 (2007)
190. S. Zorba, Y. Shapir, Y. Gao: Fractal-mound growth of pentacene thin films. Phys. Rev. B **74** 245410 (2006)
191. J.-K. Zuo, J.F. Wendelken: Evolution of mound morphology in reversible homoepitaxy on Cu(100). Phys. Rev. Lett. **78**, 2791 (1997)

Symbols

ξ	Lateral Correlation Length
λ	Wavelength
w	Interface Width (Root-Mean-Square (RMS) Roughness)
$\overline{h}$	Mean Height
α	Roughness Exponent (Tilt of Surface Normal in Chap. 9)
β	Growth Exponent (Tilt of Columnar Growth in Chap. 9)
z	Dynamic Exponent (Aggregate Height in Chap. 9)
p	Wavelength Exponent (Aggregate Growth Exponent in Chap. 9)
$\eta(\mathbf{x}, t)$	Random Noise
$P(\mathbf{k}, t)$	Power Spectral Density Function (PSD)
$R(\mathbf{r}, t)$	Autocorrelation Function
$H(\mathbf{r}, t)$	Height–Height Correlation Function
k_m	Peak PSD Position
$J_n(x)$	nth Order Bessel Function of the First Kind
$I_0(x)$	Zeroth-Order Modified Bessel Function of the First Kind
$K_\alpha(x)$	α-order Modified Bessel Function of the Second Kind
$\delta(x)$	Delta Function
$\Gamma(x)$	Gamma Function
$\langle f(\mathbf{x}, t)\rangle$	Average Value of $f(\mathbf{x}, t)$
$\sim$	Similar Statistical Behavior
$\mathcal{F}$	Fourier Transform
$\mathcal{H}$	Hankel Transform
$\mathcal{P}$	Probability

Index

Springer Series in
MATERIALS SCIENCE

Editors: R. Hull R. M. Osgood, Jr. J. Parisi H. Warlimont

50 **High-Resolution Imaging and Spectrometry of Materials**
Editors: F. Ernst and M. Rühle

51 **Point Defects in Semiconductors and Insulators**
Determination of Atomic and Electronic Structure from Paramagnetic Hyperfine Interactions
By J.-M. Spaeth and H. Overhof

52 **Polymer Films with Embedded Metal Nanoparticles**
By A. Heilmann

53 **Nanocrystalline Ceramics**
Synthesis and Structure
By M. Winterer

54 **Electronic Structure and Magnetism of Complex Materials**
Editors: D.J. Singh and D. A. Papaconstantopoulos

55 **Quasicrystals**
An Introduction to Structure, Physical Properties and Applications
Editors: J.-B. Suck, M. Schreiber, and P. Häussler

56 **SiO_2 in Si Microdevices**
By M. Itsumi

57 **Radiation Effects in Advanced Semiconductor Materials and Devices**
By C. Claeys and E. Simoen

58 **Functional Thin Films and Functional Materials**
New Concepts and Technologies
Editor: D. Shi

59 **Dielectric Properties of Porous Media**
By S.O. Gladkov

60 **Organic Photovoltaics**
Concepts and Realization
Editors: C. Brabec, V. Dyakonov, J. Parisi and N. Sariciftci

61 **Fatigue in Ferroelectric Ceramics and Related Issues**
By D.C. Lupascu

62 **Epitaxy**
Physical Principles and Technical Implementation
By M.A. Herman, W. Richter, and H. Sitter

63 **Fundamentals of Ion-Irradiated Polymers**
By D. Fink

64 **Morphology Control of Materials and Nanoparticles**
Advanced Materials Processing and Characterization
Editors: Y. Waseda and A. Muramatsu

65 **Transport Processes in Ion-Irradiated Polymers**
By D. Fink

66 **Multiphased Ceramic Materials**
Processing and Potential
Editors: W.-H. Tuan and J.-K. Guo

67 **Nondestructive Materials Characterization**
With Applications to Aerospace Materials
Editors: N.G.H. Meyendorf, P.B. Nagy, and S.I. Rokhlin

68 **Diffraction Analysis of the Microstructure of Materials**
Editors: E.J. Mittemeijer and P. Scardi

69 **Chemical-Mechanical Planarization of Semiconductor Materials**
Editor: M.R. Oliver

70 **Applications of the Isotopic Effect in Solids**
By V.G. Plekhanov

71 **Dissipative Phenomena in Condensed Matter**
Some Applications
By S. Dattagupta and S. Puri

72 **Predictive Simulation of Semiconductor Processing**
Status and Challenges
Editors: J. Dabrowski and E.R. Weber

73 **SiC Power Materials**
Devices and Applications
Editor: Z.C. Feng

Springer Series in
MATERIALS SCIENCE

74 **Plastic Deformation in Nanocrystalline Materials**
By M.Yu. Gutkin and I.A. Ovid'ko

75 **Wafer Bonding**
Applications and Technology
Editors: M. Alexe and U. Gösele

76 **Spirally Anisotropic Composites**
By G.E. Freger, V.N. Kestelman, and D.G. Freger

77 **Impurities Confined in Quantum Structures**
By P.O. Holtz and Q.X. Zhao

78 **Macromolecular Nanostructured Materials**
Editors: N. Ueyama and A. Harada

79 **Magnetism and Structure in Functional Materials**
Editors: A. Planes, L. Mañosa, and A. Saxena

80 **Micro- and Macro-Properties of Solids**
Thermal, Mechanical and Dielectric Properties
By D.B. Sirdeshmukh, L. Sirdeshmukh, and K.G. Subhadra

81 **Metallopolymer Nanocomposites**
By A.D. Pomogailo and V.N. Kestelman

82 **Plastics for Corrosion Inhibition**
By V.A. Goldade, L.S. Pinchuk, A.V. Makarevich and V.N. Kestelman

83 **Spectroscopic Properties of Rare Earths in Optical Materials**
Editors: G. Liu and B. Jacquier

84 **Hartree–Fock–Slater Method for Materials Science**
The DV–X Alpha Method for Design and Characterization of Materials
Editors: H. Adachi, T. Mukoyama, and J. Kawai

85 **Lifetime Spectroscopy**
A Method of Defect Characterization in Silicon for Photovoltaic Applications
By S. Rein

86 **Wide-Gap Chalcopyrites**
Editors: S. Siebentritt and U. Rau

87 **Micro- and Nanostructured Glasses**
By D. Hülsenberg and A. Harnisch

88 **Introduction to Wave Scattering, Localization and Mesoscopic Phenomena**
By P. Sheng

89 **Magneto-Science**
Magnetic Field Effects on Materials: Fundamentals and Applications
Editors: M. Yamaguchi and Y. Tanimoto

90 **Internal Friction in Metallic Materials**
A Reference Book
By M.S. Blanter, I.S. Golovin, H. Neuhäuser, and H.-R. Sinning

91 **Time-dependent Mechanical Properties of Solid Bodies**
By W. Gräfe

92 **Solder Joint Technology**
Materials, Properties, and Reliability
By K.-N. Tu

93 **Materials for Tomorrow**
Theory, Experiments and Modelling
Editors: S. Gemming, M. Schreiber and J.-B. Suck

94 **Magnetic Nanostructures**
Editors: B. Aktas, L. Tagirov, and F. Mikailov

95 **Nanocrystals and Their Mesoscopic Organization**
By C.N.R. Rao, P.J. Thomas and G.U. Kulkarni

96 **Gallium Nitride Electronics**
By R. Quay

97 **Multifunctional Barriers for Flexible Structure**
Textile, Leather and Paper
Editors: S. Duquesne, C. Magniez, and G. Camino

98 **Physics of Negative Refraction and Negative Index Materials**
Optical and Electronic Aspects and Diversified Approaches
Editors: C.M. Krowne and Y. Zhang

GPSR Compliance

The European Union's (EU) General Product Safety Regulation (GPSR) is a set of rules that requires consumer products to be safe and our obligations to ensure this.

If you have any concerns about our products, you can contact us on ProductSafety@springernature.com

In case Publisher is established outside the EU, the EU authorized representative is:

Springer Nature Customer Service Center GmbH
Europaplatz 3
69115 Heidelberg, Germany

Batch number: 10755943

Printed by Printforce, the Netherlands